INTRODUCTION
TO
MODERN
ELECTRONICS

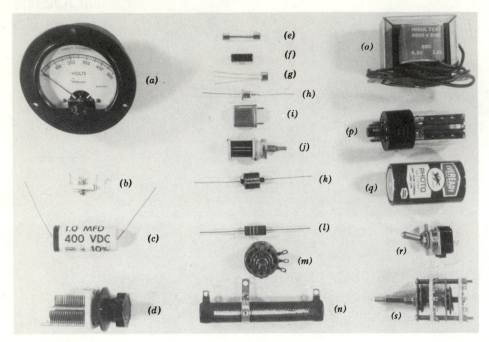

Some of the electronic components discussed in the text. Top to bottom, left to right: (*a*) dc voltmeter, (*b*) capacitor (trimmer), (*c*) capacitor (fixed), (*d*) capacitor (variable), (*e*) fuse, (*f*) integrated circuit, (*g*) transistor, (*h*) diode, (*i*) quartz crystal, (*j*) inductor (variable), (*k*) inductor (fixed), (*l*) resistor (fixed), (*m*) potentiometer, (*n*) resistor (variable), (*o*) transformer, (*p*) vacuum tube, (*q*) battery, (*r*) switch (toggle), (*s*) switch (rotary).

INTRODUCTION TO MODERN ELECTRONICS

Julien C. Sprott

The University of Wisconsin, Madison

John Wiley & Sons

New York   Chichester   Brisbane   Toronto

The cover photograph has been
reproduced by permission © 1979 Zilog, Inc.

Library of Congress Cataloging in Publication Data

Sprott, Julien C.
Introduction to modern electronics.

  Bibliography: p.
  Includes index.
  1. Electronics.   2. Electronic circuits.
  I. Title.
TK7815.S72   621.381   80-25366
ISBN 0-471-05840-8

Printed in the United States of America

10 9 8 7 6 5 4 3 2 1

Preface

One of the pleasures of teaching electronics is that few people doubt its importance and usefulness. We live in a society replete with radios, televisions, computers, and a plethora of other electronic gadgets. A knowledge of electronics is vital to a variety of occupations and useful to many more. No truly educated person can completely neglect the study of electronics.

This book came about as a result of a course taught in the Physics Department at the University of Wisconsin in Madison. The students in the course are mostly Juniors majoring in physics or engineering, and it is assumed they have had a good introductory physics course and a course in calculus. No previous knowledge of electronics is assumed.

The more advanced mathematical techniques (differential equations, complex variables, and Fourier analysis) are explained in some detail where they are first encountered, and students with no previous exposure to these topics should be able to understand them without great difficulty.

The book divides naturally into two parts. The first part (Chapters 1 to 5) covers linear circuits. The second part (Chapters 6 to 12) covers nonlinear circuits.

The flavor of the text changes somewhat after Chapter 5, moving from a careful pedogogical treatment of linear circuit analysis techniques to a broad survey of some of the more important applications of the various nonlinear components. The goal is to make the student comfortable with the relatively straightforward analysis of linear circuits before launching into the more empirical, but more interesting and useful, aspects of nonlinear circuits. A too hasty treatment of linear circuits seems to be a considerable handicap to students as they advance to the more difficult topics.

The first two chapters cover the fundamentals of direct-current (dc) circuits. Chapter 3 introduces the basic linear alternating-current (ac) components — the capacitor and the inductor — and then discusses transient circuits (i.e., circuits in which the sources are dc but are turned on or off abruptly). The equations describing such circuits are the simplest type of differential equations, and even students with no previous exposure to differential equations are able to learn to quickly solve such problems. Chapter 4 deals with sinusoidal ac circuits. The time-domain solution of one simple case is presented using differential equations, but the student is quickly introduced to the use of complex impedance as a shortcut to reduce such ac circuits to circuits for which the techniques of dc circuit analysis can be used. Chapter 5 concludes the discussion of linear circuits with an introduction to Fourier methods and distributed circuits, with emphasis on transmission lines and waveguides. This chapter is mathematically the most difficult of the book, requiring integration of complex variables, but the emphasis is on the physical ideas rather than the mathematical methods.

In Chapter 6 the diode and its applications to rectifier circuits are covered. Chapters 7 and 8 deal with the basic nonlinear active components — vacuum tubes, field effect transistors, and bipolar transistors. The emphasis on vacuum tubes runs counter to the trend in modern electronics books, but it reflects my feeling that their operation is easier for the students to understand than semiconductors. Furthermore, vacuum tubes are far from being replaced in devices such as oscilloscopes, televisions, and high-power radio transmitters In Chapter 9 the operational amplifier is discussed, which is rapidly becoming the workhorse of analog electronics. The discussion of operational amplifiers is used as an opportunity to treat the important practical subjects of negative feedback, gain-bandwidth product, noise, and circuit isolation. Chapter 10 contains a collection of other useful nonlinear devices and circuits that the student is likely to encounter. Chapter 11 deals with digital and logic circuits that form the building blocks of digital computers. To treat this important and rapidly developing field in the depth that it deserves would have required a book considerably longer than I was willing to write. Consequently, this chapter should be viewed as the barest introduction to digital electronics. Most electronics textbooks conclude with a discussion of the digital computer as the most sophisticated application of electronics, and rightly so. It seemed a shame, though, for the student to complete a course such as this without knowing how a radio or television works. Consequently, a brief chapter at the end describes communications systems, including radar, in a very general way.

The text was originally intended to contain just enough information for the average student to absorb comfortably in a one-semester course. During subsequent revisions, topics were added here and there, and a certain selectiveness would probably now be required. The student might be encouraged to read the entire book, but certain sections, such as the whole of Chapters 5 and 12 and most topics in Chapter 10, could be touched on lightly or not at all without great loss of continuity.

Problems are an integral part of learning any technical subject such as electronics, and each chapter contains a number of problems designed to test and in many cases expand the student's knowledge of the subject. Problems in which the student can find the right equation and just plug in the numbers are largely avoided. The problems span a considerable range of difficulty. Answers to the odd-numbered problems are included in Appendix K, and a Solutions Manual for the text is available for instructors.

The course is designed to be accompanied by a laboratory. Although electronics can be taught strictly from a book, there is no substitute for the kind of hands-on experience that a laboratory provides. At this level, the equipment required is very modest and represents an excellent investment in the quality of teaching. Appendix L contains a number of suggested laboratory experiments that have been used for many years in my course at the University of Wisconsin.

I am indebted to Professor Stewart Prager for a careful reading of an early version of the manuscript and for numerous helpful suggestions. Additional suggestions were provided by Tom Lovell, Kevin Miller, Don Holly, and Mike Zarnstorff. The laboratory experiments were in large part inherited from Professor Wilmer Anderson. The tedious task of typing the many revisions of the manuscript was capably performed by Mike Seldomridge and Kay Shatrawka. Finally, to my students, who have taught me more about electronics than they realize, I want to dedicate this book.

Madison Wisconsin   **J. C. Sprott**
April, 1981

Contents

INTRODUCTION TO MODERN ELECTRONICS

dc Circuit
Components

1.1 Current and Voltage

The study of electronics is largely a study of the behavior and relationship of two quantities — **current** and **voltage**. It is crucial, therefore, that the meanings of current and voltage be clearly understood.

Current is defined as the amount of electrical charge crossing a surface per unit time. In the International System, abbreviated as SI, from the French (formerly MKS), the unit of current is called the **ampere** (abbreviated amp or A) and is equal to 1 coulomb per second:

$$1 \text{ A} = 1 \text{ C/s}$$

Electrical currents are usually carried by electrons. By historical accident, the charge of the electron $(-e)$ was defined as being negative and is given by

$$e = 1.6 \times 10^{-19} \text{ C}$$

Therefore, the electrical current always flows in a direction opposite to the direction in which the electrons move. In the study of electronics, one rarely worries about what electrons are doing, and it is usually more convenient to imagine positive charges which flow in the same direction as the current. A current of 1 A requires 6.2×10^{18} electrons to cross a surface during each second. Since this number is so large, we normally don't notice the quantization of currents.

Voltage is defined as the amount of energy required to move a unit of electrical charge from one place to another. In the SI system, the unit of voltage is called the **volt** (abbreviated V) and is equal to one joule per coulomb:

$$1 \text{ V} = 1 \text{ J/C}$$

The use of "V" for both an algebraic quantity and a unit of voltage is a potential source of confusion in expressions such as $V = 5$ V. (Remember: the algebraic quantity is in italic.) Implicit in the definition is the fact that only voltage *differences* have meaning. To say that a certain point in an electrical circuit has a voltage of 12 V is meaningless unless one designates a point in the circuit as a reference (zero voltage) point. Often such a reference point is referred to as a **ground**, because it is usually connected to a metal case or chassis which encloses the circuit and is in turn connected electrically to the earth. The symbol for a connection to ground is $\perp$. Circuits are often grounded as a safety precaution to ensure that those parts of a circuit that the user is likely to touch are at the same voltage as the ground on which he or she is standing.

Electrical energy has to be supplied to move a positive charge (to make a current flow) toward a point with a higher (more positive) voltage. Current will tend to flow of its own accord from a high voltage to a lower voltage point, and in so doing will dissipate (convert into other forms) electrical energy. Other terms which are often used instead of voltage are **potential** and **electromotive force (emf)**.

A common analogy which is useful for visualizing the concepts of voltage and current is a water system in which a pump raises a mass m of water to a water tower at height h above the ground from which it flows back to ground level. The energy required to raise the water to the tower is given by mgh where g is the acceleration due to gravity ($g = 9.8$ m/s^2). The gravitational potential (potential energy per unit mass) is proportional to the height above the ground and is analogous to the voltage in an electrical circuit. The flow rate, kilograms per second, is analogous to the current in an electrical circuit.

1.2 Resistance

If a rod of material has a voltage difference V between its ends, a current I will flow through the rod. The current will flow from the high voltage end toward the end at lower voltage. To a very good approximation, the voltage and current are proportional:

$$V = IR \tag{1.1}$$

where the proportionality constant R is called the **resistance**. Equation 1.1 is called **Ohm's law**, although on close examination it is never precisely obeyed. The resistance R has units of volts per amp which we call an **ohm** (abbreviated Ω):

$$1 \ \Omega = 1 \ \text{V/A}$$

Typical resistor values range from about 1 Ω to about 1 MΩ (equal to 10^6 ohms). A related quantity is the **conductance** G, defined as $1/R$. The unit of conductance is the **siemens** (formerly called **mho**) and is abbreviated $\mho$.

There is a vast difference between the resistance of different materials. Materials with very low resistance such as silver, copper, aluminum, and other metals are called **conductors**. Materials with very high resistance such as glass, rubber, and air are called **insulators**. A perfect or ideal conductor is one in which $R = 0$, and so according to equation 1.1, an ideal conductor cannot have any voltage difference between its ends, even though a large current may be flowing through it. Certain metal alloys at temperatures near absolute zero ($-273°$C) are ideal conductors and are called **superconductors**. A perfect or ideal insulator is one in which R is infinite (or $G = 0$);

hence no current will flow, even though a large voltage difference may exist. The only ideal insulator is a perfect vacuum.

Some materials such as carbon have medium resistance and can be used to form electrical circuit elements called **resistors**. Composition resistors are produced by mixing small grains of carbon in an insulating resin and molding the composite into a short cylinder with conducting copper leads attached to each end. These leads are called **terminals**. By varying the amount of carbon, resistance values ranging from a few ohms to many megohms can be made. Resistors with resistance less than about 100 kΩ can be made using a coil of wire of some poorly conducting alloy such as manganin (copper, manganese, and nickel). Such wire-wound resistors are more expensive, but they have higher stability and heat dissipation capabilities. The resistance value is commonly indicated by a color code as described in figure 1.1.

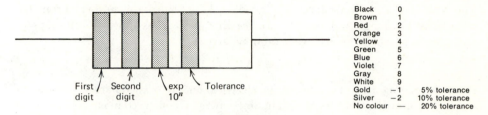

Black	0	
Brown	1	
Red	2	
Orange	3	
Yellow	4	
Green	5	
Blue	6	
Violet	7	
Gray	8	
White	9	
Gold	−1	5% tolerance
Silver	−2	10% tolerance
No colour	—	20% tolerance

First digit / Second digit / exp 10^n / Tolerance

Fig. 1.1 Resistor color code. The third band indicates the power of 10 by which the first two digits are multiplied. A resistor with yellow-violet-red-silver would be 4700 Ω ± 10%.

Real resistors don't precisely obey Ohm's law. When current passes through a resistor, it gets hot, and its resistance changes slightly. Also, the current does not change instantly when the voltage is abruptly changed, and vice versa. It is, nevertheless, useful to define an **ideal resistor** as one in which Ohm's law is exactly obeyed. An electrical circuit containing such ideal components can be analyzed in a systematic manner and usually behaves in a way that adequately approximates the behavior of the corresponding circuit with real components.

The symbols used to represent ideal resistors are shown in figure 1.2. Variable

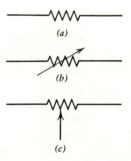

(a)

(b)

(c)

Fig. 1.2 Symbols for ideal resistors. *(a)* Fixed. *(b)* Variable. *(c)* Potentiometer.

resistors are sometimes called **potentiometers** ("pot" for short). Note that a potentiometer is a three-terminal device in which the resistance between two of the terminals is fixed, but the resistance between the third terminal and either end is variable. The volume control in a radio receiver is a potentiometer with a logarithmic **taper** (the angle of rotation is approximately proportional to the logarithm of the resistance). Most potentiometers, especially those found on scientific instruments, have a linear taper. A logarithmic taper is used for volume controls, because the ear has an approximately logarithmic response; that is, every doubling of the acoustic energy leads to a constant increment in perceived loudness.

1.3 Power

The ratio V/I is the resistance. The product VI is called the **power**. From the definitions of voltage and current, we see that power has units of volt-amps or joules/seconds and is called a **watt** (abbreviated W):

$$1 \text{ W} = 1 \text{ VA} = 1 \text{ J/s}$$

Power is the time rate of change of the electrical energy of the charged particles as they move through an electrical circuit. Since energy is conserved in nature, electrical energy can be produced only by depleting some other form of energy, such as chemical energy in a battery or mechanical energy in a generator. Electrical energy can be dissipated only by converting it to some other form such as heat, light, or sound. A resistor converts electrical energy into heat, and the rate at which this happens is the power,

$$P = IV = I^2 R = V^2/R \tag{1.2}$$

where the last two forms are derived from Ohm's law. The amount of power that a resistor can safely dissipate without overheating is dependent, among other things, on its physical size. A typical resistor with a diameter about the size of a pencil can safely dissipate about 1 W of power.

1.4 Sources

Before proceeding further with our study of electronics we must consider sources of electrical power and their properties. We have already mentioned batteries that store energy in chemical form and convert it into usable electrical energy on demand and generators that convert mechanical energy into electrical energy. Other examples are solar cells that convert light into electricity, thermocouples that convert heat into electricity, and microphones that convert sound into electricity. The most common type of source that is encountered is the power supply which converts one form of electrical energy into another. These sources often have a rather complicated relationship between the voltage and current they are able to supply. Therefore, it is useful to consider ideal sources which have well-prescribed mathematical properties.

These ideal sources represent limiting cases of real sources and are called the **voltage source** and the **current source**. The symbols we will use for these sources are shown in figure 1.3. The ideal voltage source has the property that the voltage across its

Fig. 1.3 (*a*) Ideal voltage source. (*b*) Ideal current source.

terminals is constant (V) no matter what current flows through it. The ideal current source has the property that the current through it is constant (I) no matter what voltage appears between its terminals. In the ideal voltage source, the symbols $+$ and $-$ only indicate which terminal is at the higher voltage. Since any reference can be chosen as zero volts, we don't know whether a terminal is positive or negative until we examine the circuit to which the source is connected and determine what point in the circuit is being used as the reference (ground).

A battery is a reasonable approximation to a voltage source, and, in fact, batteries are sold according to their voltage rating (1.5, 6, 9, and 12 V are common examples). But in reality, the voltage across the terminals of a battery will decrease when a current flows out of its positive terminal. Conversely, the voltage will increase when current flows into its positive terminal, as when an automobile battery is being charged. Since a real battery does not behave in an ideal manner, we will reserve a special symbol for it, as indicated in figure 1.4(*a*). A better approximation to a real battery is an ideal voltage source connected to an ideal resistor, as shown in figure 1.4(*b*). As a current I flows out of the positive terminal of the source, it flows through the resistor r (called the **internal resistance** or the **source resistance**)

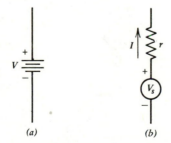

Fig. 1.4 (*a*) Battery. (*b*) Real source approximated by an ideal voltage source and an ideal resistor.

producing a voltage drop Ir in the resistor so that the voltage at the terminals of the source/resistor combination is given by

$$V = V_s - Ir \tag{1.3}$$

For $I = 0$, the full voltage V_s appears, but the voltage decreases to zero and can even go negative if a sufficiently large current is produced by other sources in the circuit. When a battery "runs down," the voltage V_s decreases, but equally important, the internal resistance increases, until, finally, no power can be delivered to the circuit to which the battery is connected. Most batteries, such as the carbon-zinc dry cell used in flashlights and portable radios, run down gradually. Mercury batteries maintain a nearly constant voltage and resistance throughout most of their life and then stop providing power very abruptly. Nickel-cadmium batteries share the same property and have the additional advantage of being rechargeable.

Equation 1.3 suggests how a reasonable approximation to a current source can be made. Solving equation 1.3 for I gives

$$I = \frac{V_s - V}{r}$$

If V_s is made very large $(V_s \gg V)$, then

$$I \simeq V_s/r$$

and the voltage source/resistor combination produces a nearly constant current of magnitude V_s/r. A real source such as a battery thus has properties intermediate between an ideal voltage source and an ideal current source. These relationships are illustrated in figure 1.5, in which we plot the voltage across a resistor, an ideal voltage

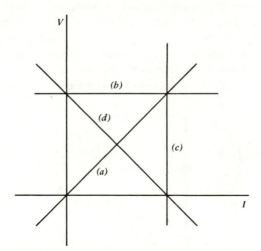

Fig. 1.5 Voltage as a function of current for (*a*) resistor, (*b*) ideal voltage source, (*c*) ideal current source, and (*d*) real source.

source, an ideal current source, and a battery as a function of the current flowing through them. Any combination of ideal resistors and ideal sources will produce a straight line on such a graph. For this reason, these components are said to be **linear**, and circuits which contain only linear components are called **linear circuits**. The equations that describe the voltage and current in such circuits will always be linear equations, and hence the solutions are straightforward. In a resistor, the current always flows from positive to negative, but in a source, the relative direction of current and voltage can be either the same or opposite, depending on the rest of the circuit. By convention, the current is considered positive when it flows out of the positive terminal of a voltage source.

We are now ready to consider the simplest possible electrical circuit, as shown in figure 1.6. It consists of an ideal voltage source connected to an ideal resistor with ideal conductors. Recall that every point on an ideal conductor is at the same voltage, and so the resistor in figure 1.6 must have a voltage V across its terminals with the

Fig. 1.6 Ideal voltage source connected to ideal resistor.

higher voltage at the top. From Ohm's law, a current $I = V/R$ must then flow downward through the resistor. For dc circuits, charge cannot accumulate at any point in the circuit, and so the same current must flow in a clockwise direction everywhere around the loop. A general feature of circuits is that currents can only flow in loops that are closed. In fact, that's why they're called circuits! Note that in this circuit, the current flows from negative to positive through the source, which will always be the case for circuits that contain a single source. But for circuits with two or more sources, one source can cause current to flow backward through another source, as is the case when a battery is being charged.

In the circuit of figure 1.6, the source produces an electrical power equal to VI. This power is transmitted to the resistor without loss. The power dissipated in the resistor is also VI or V^2/R (equation 1.2). Note that as R approaches zero, the current and power both go to infinity. A real source cannot provide infinite power, and the internal resistance causes the voltage to drop as the current increases. Similarly, an ideal current source connected to a resistor will produce a voltage $V = IR$ across a resistor, and the power (I^2R) will approach infinity as the resistor is made large. Therefore, just as it is unwise to connect the terminals of a battery (voltage source) together, it is unwise to leave the terminals of a current source unconnected. In one case, a large current flows. In the other, a large voltage develops.

Circuits are often protected from the damage that could result from large currents by means of a **fuse**. A fuse is a low-resistance conductor that melts and breaks the circuit when the current exceeds a prescribed value. It is then discarded and replaced. Fuses are commonly available with ratings ranging from a small fraction of an ampere to several hundred amperes. The symbol for a fuse is —oɅo— . A closely related device is a **circuit breaker** which mechanically breaks the circuit when the current exceeds a prescribed value. The advantage of the circuit breaker is that it can be manually reset. Large voltages can be protected against by the use of devices called **transient suppressors** or **thyrites**. These devices have very high resistance up to some value of voltage at which the resistance drops abruptly. Note that none of these protective devices is linear, but in normal use the nonlinear behavior does not occur, and they can usually be treated simply as ideal conductors or insulators.

1.5 Circuit Reduction

The analysis of an electrical circuit normally consists of determining the current and/or voltage at one or more points in the circuit. A complicated electrical circuit can often be analyzed by reducing the circuit to a simpler circuit for which the solution is known. Consider, for example, the circuit in figure 1.7. The two resistors

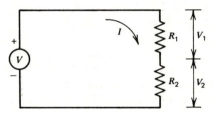

Fig. 1.7 Resistors in series add according to $R = R_1 + R_2$.

R_1 and R_2 are said to be connected in **series**. The sum of the voltage drops across each resistor must equal the voltage of the source:

$$V = V_1 + V_2$$

Since the same current I flows through each resistor, Ohm's law can be applied to each resistor to give

$$V = IR_1 + IR_2 = I(R_1 + R_2)$$

The sum $R_1 + R_2$ is called the **equivalent resistance** for resistors in series:

$$R = R_1 + R_2 \tag{1.4}$$

The circuit of figure 1.7 can thus be reduced to the simpler circuit of figure 1.6 by using the equivalent resistance. The voltage drop across one of the series resistors, say

R_1, can be determined from the current I using Ohm's law:

$$V_1 = IR_1 = \frac{R_1 V}{R} = \frac{R_1 V}{R_1 + R_2} \tag{1.5}$$

This equation is called the **voltage divider relation**, and is extremely useful. It says that for resistors in series, the voltage divides in proportion to the resistance.

Another example is shown in figure 1.8 in which the two resistors are connected

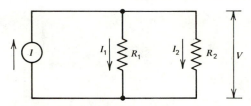

Fig. 1.8   Resistors in parallel add according to $1/R = 1/R_1 + 1/R_2$.

in **parallel**. In this case, the current I divides between the two resistors in such a way that

$$I = I_1 + I_2$$

By applying Ohm's law and using the fact that the same voltage V appears across each resistor we obtain

$$I = \frac{V}{R_1} + \frac{V}{R_2} = V\left(\frac{1}{R_1} + \frac{1}{R_2}\right)$$

The equivalent resistance for resistors in parallel is thus determined from

$$\frac{1}{R} = \frac{1}{R_1} + \frac{1}{R_2} \tag{1.6}$$

For the special case of two resistors in parallel, equation 1.6 can be written in the convenient form

$$R = \frac{R_1 R_2}{R_1 + R_2}$$

Equations 1.4 and 1.6 can be generalized to any number of resistors:

$$R = \sum_i R_i \quad \text{(series)}$$

$$\frac{1}{R} = \sum_i \frac{1}{R_i} \quad \text{(parallel)}$$

The current through one of the resistors, say R_1, can be determined from the voltage

V using Ohm's law:

$$I_1 = \frac{V}{R_1} = \frac{IR}{R_1} = \frac{R_2 I}{R_1 + R_2} \tag{1.7}$$

This equation is called the **current divider relation**. Note the similarity of equation 1.7 and equation 1.5, but notice that in the current divider, the current in resistor 1 is proportional to R_2 rather than to R_1, as was the case with the voltage divider.

Sources can also be connected in series or parallel as indicated in figure 1.9.

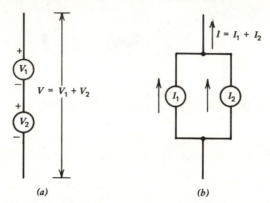

Fig. 1.9 (*a*) Voltage sources in series add. (*b*) Current sources in parallel add.

Voltage sources in series add according to

$$V = V_1 + V_2$$

Current sources in parallel add according to

$$I = I_1 + I_2$$

Ideal voltage sources cannot be connected in parallel unless they have the same voltage. Similarly, ideal current sources cannot be connected in series unless they have the same current.

Sources can also be connected in series or parallel with resistors. The case of a resistor in series with a voltage source has already been considered in section 1.4. A resistor in parallel with a voltage source has no effect on the circuit to which it is connected, since the voltage produced by an ideal voltage source is independent of the current drawn from the source. For a similar reason, a resistor in series with a current source has no effect on the rest of the circuit.

The case of a resistor in parallel with a current source is more interesting, however. In figure 1.10, a current equal to $I_s - I$ must flow downward through resistor r, so that the voltage V is

$$V = (I_s - I)r$$

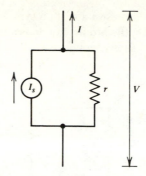

Fig. 1.10 A resistor in parallel with a current source is undistinguishable from the same resistor in series with a voltage source $V_s = I_s r$.

This equation is identical to equation 1.3, provided we set

$$V_s = I_s r$$

with r the same in the two circuits. The circuit in figure 1.10 is therefore indistinguishable from the circuit in figure 1.4(b) if the above conditions are satisfied. The ability to switch back and forth between these two representations is a powerful tool for circuit analysis by successive reduction. For example, the circuit in figure 1.11(a) can be reduced to the circuit in figure 1.11(b) which is just a voltage divider.

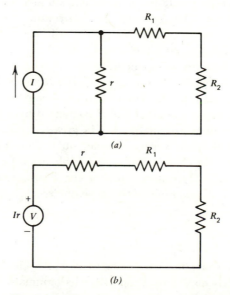

Fig. 1.11 The circuit in (a) is equivalent to that in (b), which is just a voltage divider.

Occasionally, a circuit will possess a certain symmetry that greatly simplifies its analysis. Consider, for example, the problem of determining the equivalent resistance of the combination of resistors shown in figure 1.12. Since no two resistors are either

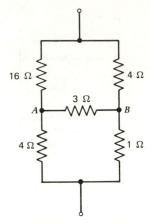

Fig. 1.12 Example of a circuit whose symmetry can be exploited to yield an equivalent resistance of 4 Ω.

in series or parallel, circuit reduction cannot be applied. Imagine, however, that the 3-Ω resistor were missing. Then the circuit becomes two voltage dividers, and any voltage applied between the two terminals will divide in such a way that points A and B are at the same voltage. Consequently, the 3-Ω resistor can be put back in the circuit without having any effect, since no current will flow through it. In fact, the equivalent resistance of the circuit is the same if the 3-Ω resistor were replaced with a resistor of any value. As an exercise, one might verify by circuit reduction that the equivalent resistance of the circuit in figure 1.12 is 4 Ω either with the 3-Ω resistor replaced by an infinite resistance (no connection between points A and B) or with it replaced by a zero resistance (points A and B connected by a perfect conductor).

Another type of circuit whose analysis can be greatly simplified is one in which a given pattern of circuit elements is repeated indefinitely. One link in the chain can be removed and the remaining chain is indistinguishable from the original chain. This analysis technique is actually quite useful for the distributed circuits which are described in Chapter 5. An example of this type of circuit is given in problem 1.9.

1.6   Meters

Meters are devices that measure current or voltage. An ideal **ammeter** has zero resistance and hence has no voltage drop across its terminals. An ideal **voltmeter** has infinite resistance and hence draws no current from the circuit to which it is connected. The ammeter produces a reading proportional to the current through its terminals. The voltmeter produces a reading proportional to the voltage across its terminals. The symbols for ideal meters are shown in figure 1.13. An ideal ammeter can be connected in series with a circuit without disturbing it. An ideal voltmeter can

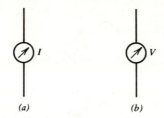

(a)   (b)

Fig. 1.13   (a) Ideal ammeter. (b) Ideal voltmeter.

be connected in parallel with a circuit without disturbing it. An ideal meter consumes no power, since either I or V is always zero.

Just as real sources are intermediate between ideal voltage and current sources, real meters are intermediate between ideal ammeters and ideal voltmeters. A common type of meter is the **D'Arsonval galvanometer**, in which a current through an electromagnet produces a torque that rotates a spring-loaded needle through an angle that is proportional to the current. The coil of the electromagnet has some resistance, and hence a small voltage drop occurs when current flows. Figure 1.14 shows two equivalent representations of a real meter in terms of ideal

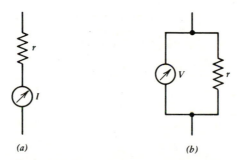

(a)   (b)

Fig. 1.14   Equivalent representations of real meters. (a) Ammeter. (b) Voltmeter.

circuit components. Voltmeters are usually made by placing a large resistor r in series with a galvanometer, as shown in figure 1.14(a). In such a case, the ammeter will read a current $I = V/r$, and it can be labeled to indicate the voltage V. A real meter always consumes power, since the product VI is never exactly zero.

The **sensitivity** of a galvanometer is expressed in terms of the current required to produce a full-scale reading. A large sensitivity means a small current is required, and so the sensitivity is defined as the inverse of the current required for a full-scale reading. From Ohm's law, the units of inverse current are ohms per volt. The sensitivity expressed in this way is a very useful number, because it tells how much resistance must be placed in series with a meter to use it as a voltmeter. For example, a typical galvanometer has a sensitivity of 20,000 Ω/V and an internal resistance of 5000 Ω. In order to make such a meter read 10 V full scale, a total series resistance of 200 kΩ is required. But the galvanometer already has 5000 Ω, and so an additional resistance of 195 kΩ should be placed in series with the galvanometer.

A galvanometer can also be used to measure currents larger than would normally produce a full-scale reading. This is done by placing a resistor (called a **shunt**) in parallel with the meter to form a current divider, as shown in figure 1.15. If the

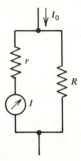

Fig. 1.15 A shunt (R) placed in parallel with a galvanometer permits a large current (I_0) to be measured.

galvanometer has a sensitivity of 20,000 Ω/V, the meter reads full scale when $I = 50\ \mu$A. If we want the meter to read full scale when $I_0 = 1$ A, the current divider relation (equation 1.7) requires that R be chosen so that

$$I = \frac{I_0 R}{R + r}$$

For $r = 5000\ \Omega$, the solution is

$$R = \frac{Ir}{I_0 - I} \simeq 0.25\ \Omega$$

Perhaps the most useful piece of electronic test equipment is the **multimeter** or **volt-ohm-milliammeter** (VOM), which typically consists of a sensitive galvanometer labeled with numerous scales and a variety of resistors that can be inserted in series or parallel by a range setting on the instrument in order to permit various full scale readings for voltage and current. Several ohm scales are also usually provided for measuring resistors (see problem 1.20). More sophisticated instruments incorporate a vacuum tube (**vacuum tube voltmeter**, or VTVM) or an FET (see Chapter 7) so that the meter is more nearly ideal. A type of vacuum tube or FET voltmeter with an exceedingly high internal resistance (often $\gtrsim 10^{14}\ \Omega$) is called an **electrometer**. A particularly convenient device is the **digital voltmeter** (DVM) or

digital multimeter (DMM) in which the voltage and other quantities are displayed directly as a number, usually with three or four digits.

A real meter always perturbs the current or voltage that is being measured. The more nearly ideal the meter, the less the perturbation. This does not mean that accurate measurements cannot be made with a real meter. If the internal resistance r of the meter is known, the effect of the meter can be taken into account, and the voltage or current in the absence of the meter can be inferred by analysis of the circuit to which it is connected.

The quality of a meter is specified by three independent parameters. The internal resistance determines how nearly ideal the meter is and hence how much it perturbs the circuit being measured. The sensitivity is a measure of the current required to produce a full-scale reading. The **accuracy** is a measure of how nearly the meter reading corresponds to the actual current or voltage at its terminals. A meter that is accurate at one point on its scale but inaccurate at another point is said to lack **linearity**. An ideal meter can be quite inaccurate, especially if it has been abused. Similarly, a meter can be perfectly accurate but quite nonideal.

Finally, note that an ideal ammeter connected in parallel with an ideal voltage source is a contradiction. Similarly, an ideal voltmeter cannot be connected in parallel with an ideal current source. Any attempt to make such a connection with real sources and meters will likely result in damage to the meter, the source, or both.

1.7 Summary

In this chapter the fundamental concepts of current and voltage have been defined. The ratio V/I is the resistance. The product VI is the power. The five basic, linear, ideal, dc circuit components have been introduced: resistor, voltage source, current source, ammeter, and voltmeter. The technique of analyzing circuits by circuit reduction has been introduced. The rules for applying this technique are summarized in figure 1.16.

The perceptive reader will notice that voltage and current play a symmetric role in many of the discussions in this chapter, as do series and parallel. Any statement in this chapter that contains the words "voltage" and "series" will also be true if "voltage" is replaced with "current" and "series" with "parallel," and vice versa. These terms are called **conjugate pairs**. Many more such conjugate pairs are introduced in the following chapters. The reader should be alert for these, as they simplify the study of electronics and add to its beauty.

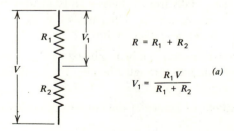

$$R = R_1 + R_2$$

$$V_1 = \frac{R_1 V}{R_1 + R_2} \qquad (a)$$

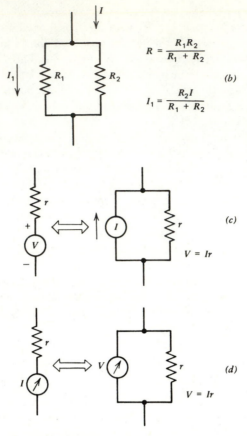

$$R = \frac{R_1 R_2}{R_1 + R_2}$$

(b)

$$I_1 = \frac{R_2 I}{R_1 + R_2}$$

(c)

$$V = Ir$$

(d)

$$V = Ir$$

Fig. 1.16 Summary of circuit reduction rules.

Problems

1.1 Calculate the average velocity of the electrons in a copper wire of 1 mm diameter carrying a current of 1 A. The density of free electrons in copper is 8.5 $\times 10^{28}$ electrons/m^3.

1.2 If 1000 kg of water falls from a height of 10 m and its potential energy is converted to electrical energy with an efficiency of 10%, how many coulombs of charge can be raised through a voltage of 100 V?

1.3 How much electrical power can be produced by a hydroelectric power plant if 1000 kg of water per second falls through a height of 10 m and its energy is converted to electricity with an efficiency of 10%?

1.4 Find the equivalent resistance of the circuit shown below:

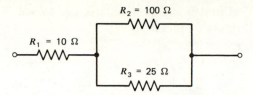

1.5 Find the equivalent resistance of the circuit shown below:

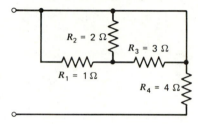

1.6 Find the equivalent resistance of the circuit shown below:

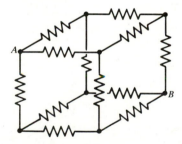

1.7 For the circuit below in which all the resistors are 1 Ω, calculate the equivalent resistance between points A and B.

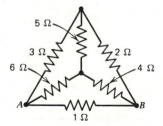

1.8 For the circuit below, calculate the equivalent resistance between points A and B.

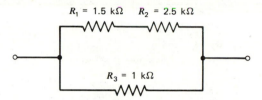

1.9 Find the equivalent resistance R of the circuit below which extends indefinitely to the right with all resistors equal to R_0:

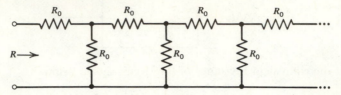

1.10 Calculate the current I_2 in the circuit below:

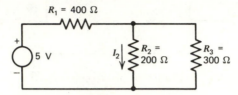

1.11 Calculate the voltage V_4 in the circuit below:

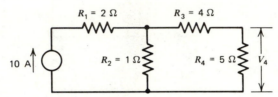

1.12 For the circuit in problem 1.11 calculate the power produced by the current source and the power dissipated by R_4.

1.13 In the circuit below, each of the resistors is rated for a maximum dissipation of 2 W. What size fuse should be used to protect the resistors from damage?

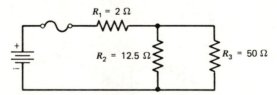

1.14 Calculate the voltage V_2 in the circuit below:

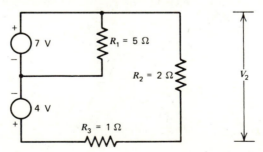

1.15 Calculate the voltage V_4 in the circuit below:

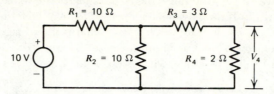

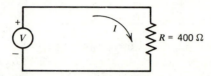

1.16 In the circuit below, the current I is measured with a real ammeter (not shown) with a 100-Ω internal resistance. If the meter reads 0.5 A, what is the current in the absence of the meter?

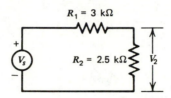

1.17 In the circuit below, the voltage V_2 is measured with a 10-V full-scale voltmeter that consists of a resistor in series with a 1000 Ω/V galvanometer. What voltage V_s is required to make the voltmeter read 5 V?

1.18 In each of the circuits below, the voltmeters are non-ideal, with a sensitivity of 20,000 Ω per volt and read 8 V on the 10-V scale, and the ammeters are nonideal and have internal resistances of 5000 Ω and read 160 μA. Calculate the values of R_1 and R_2.

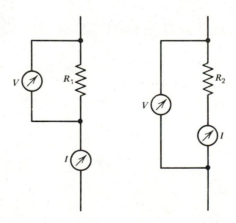

1.19   A galvanometer with $20{,}000\,\Omega$ per volt sensitivity and $5000\,\Omega$ internal resistance is used with an appropriate series or parallel resistor to make measurements in the circuit below. (a) If the galvanometer is used as a voltmeter to measure V_2 and the meter reads 0.5 V on a 1-V full-scale range, what is the voltage V of the source? (b) If the galvanometer is used as an ammeter to measure I and the meter reads $50\ \mu A$ on a $100\text{-}\mu A$ full-scale range, what is the voltage V of the source?

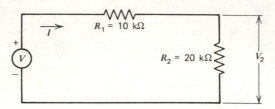

1.20   The circuit below is commonly used as part of a VOM to measure an unknown resistance R. The procedure is to first set $R = 0$ and adjust R_2 for full-scale deflection of the meter $(50\ \mu A)$. Then the unknown R is placed in the circuit and the current is measured without disturbing R_2. Show that the reading on the meter is independent of the voltage V so that a battery of unknown condition can be used without losing accuracy. What value of R will give a meter deflection of $1/2$ scale if $R_2 \gg R_1$? Repeat for $1/10$ scale and $9/10$ scale.

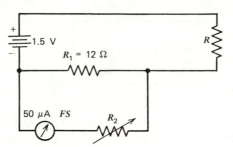

1.21   An ammeter with a 1-A full-scale range and an internal resistance of $0.1\,\Omega$ is connected in parallel with a 12-V battery that has an internal resistance of $0.2\,\Omega$. Calculate the current that flows through the ammeter, and describe what is likely to happen to the meter and to the battery.

1.22   The circuit below is called an **Ayrton shunt**. If the ammeter is ideal with a full-scale reading of $1\ \mu A$, what current must flow between terminal 1 and ground to produce a full-scale reading of the meter? Repeat for terminals 2, 3, and 4.

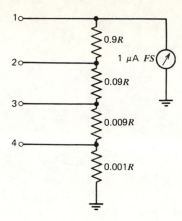

1.23 Real meters can be connected in series or parallel to permit readings outside the normal range of the meters. Suppose you had two VOM's, each capable of reading up to 1000 V and 1 A. If meter 1 has a 20,000 Ω/V galvanometer and meter 2 has a 5000 Ω/V galvanometer, show how you would connect the meters to read a voltage of 1100 V, and calculate the reading of each meter.

Circuit
Theorems

2.1 Kirchhoff's Laws

As useful as the circuit-reduction techniques of the last chapter are, some circuits require a more sophisticated analysis. For example, the circuit shown in figure 2.1

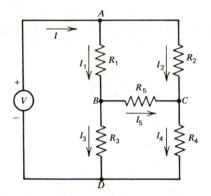

Fig. 2.1 The Wheatstone bridge is an example of a circuit that cannot be analyzed by simple circuit reduction.

cannot be analyzed by simple circuit reduction. This circuit is called a **Wheatstone bridge**, and it has a number of important applications which will be discussed later. The most general method for analyzing circuits makes use of **Kirchhoff's laws**. The use of Kirchhoff's laws has the virtue that it always works whether the circuit contains linear or nonlinear elements, no matter how complicated. Kirchhoff's laws form the basis for all the theorems to be discussed in this chapter.

Before stating Kirchhoff's laws, it is useful to define certain terms:

Node: a point where three or more circuit elements are connected together.

Branch: a circuit element or series of elements that connect two adjacent nodes.

Loop: a circuit path that begins at a node, passes through one or more nodes, and ends at the same node at which it started.

Mesh: a loop that does not contain any branches in its interior (also called an **elementary loop**).

These concepts are most easily understood by referring to figure 2.1. This circuit has four nodes labeled A, B, C, and D. It has six branches, five containing resistors and one containing a voltage source. There are a total of seven loops, but only three of these are meshes (ABC, BCD, and ABD). With these terms clearly in mind, we can state Kirchhoff's laws, which follow directly from the definitions of current and voltage:

Kirchhoff's current law: The sum of the currents flowing into a node is zero.

Kirchhoff's voltage law: The sum of voltage drops around a loop is zero.

Kirchhoff's current law is a statement of the conservation of electric charge, because if the total current flowing into a node were other than zero, an infinite charge would eventually build up at the node. Kirchhoff's voltage law is a statement of the conservation of energy, because if a charge moving around a loop came back to its starting point at a voltage different from what it had initially, its energy would have changed.

For the example in figure 2.1, Kirchhoff's current law yields four equations, one for each node:

$$I = I_1 + I_2$$
$$I_1 = I_3 + I_5$$
$$I_2 + I_5 = I_4$$
$$I_3 + I_4 = I$$

These equations are not all independent, however. The fourth equation can be derived by substituting I_1 and I_2 from the middle two equations into the first equation. It will always be the case that Kirchhoff's current law gives one extra equation, and so any one of the nodes can be ignored. The reason for this is that the current is constrained to flow in closed loops, so that the current flowing into any node must be zero if the current flowing into all the other nodes are zero.

Kirchhoff's voltage law yields seven equations, one for each loop, but, again, not all are independent, because many of the loops are the sum of smaller loops. In fact, examination of the circuit shows that all the loops can be constructed from various combinations of meshes, and so there are three independent loop equations, because there are three meshes:

$$I_1 R_1 + I_3 R_3 - V = 0$$
$$I_1 R_1 + I_5 R_5 - I_2 R_2 = 0$$
$$I_3 R_3 - I_4 R_4 - I_5 R_5 = 0$$

One can circle a mesh either clockwise or counterclockwise, but either way, the voltage drop is positive if one goes in the same direction as the arrow and negative if one goes opposite to the arrow. The arrows can be drawn in either direction, however. When the equations are solved, a positive solution for the current in a branch means the current flows in the direction of the arrow, and a negative solution means it flows opposite to the arrow. When one goes from negative to positive in a source, the voltage rises, and hence the voltage drop is negative, irrespective of the direction of the current.

The number of unknowns in a circuit is always equal to the number of branches, because if the current in every branch is known, all the voltages can be calculated from Ohm's law. For the case in figure 2.1, there are six branches and hence six unknowns. Kirchhoff's laws give six independent equations (three current and three voltage), and so a solution exists. Such will always be the case:

$$\text{Number of unknowns} = \text{number of branches}$$
$$= \text{number of nodes} - 1 + \text{number of meshes}$$

Six linear algebraic equations in six unknowns can be solved by several means, all of which are somewhat tedious. One way is to eliminate unknowns one at a time by solving for them in terms of the other unknowns and substituting into the remaining equations until the system of equations is reduced to a single equation in a single unknown. That unknown is then calculated, and it can be substituted into the previous equation, and so on, until all the unknowns are determined. Another way is by the use of determinants as described in virtually all elementary calculus texts. Much of the tedium of solving systems of linear algebraic equations has now been relegated to computers, and a circuit designer with access to such a computer would be well advised to take advantage of its capabilities.

A slight simplification results from using what is called the **loop current** technique. In figure 2.2, the Wheatstone bridge is redrawn, but rather than labeling

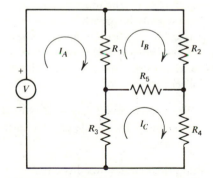

Fig. 2.2 Analysis of Wheatstone bridge using the loop current technique.

each branch current, as was done in figure 2.1, we have defined a loop current for each mesh. The current in a branch is a sum of the loop currents flowing through that branch:

$$I = I_A$$
$$I_1 = I_A - I_B$$
$$I_2 = I_B$$
$$I_3 = I_A - I_C$$
$$I_4 = I_C$$
$$I_5 = I_C - I_B$$

By defining loop currents in this way, Kirchhoff's current law is automatically satisfied, since every loop current that flows into a node also flows out of the node. The result is to reduce the problem to one of solving three equations in three unknowns. The penalty is that the equations are somewhat more complicated:

$$(I_A - I_B)R_1 + (I_A - I_C)R_3 - V = 0$$
$$(I_A - I_B)R_1 + (I_C - I_B)R_5 - I_B R_2 = 0$$
$$(I_A - I_C)R_3 - I_C R_4 - (I_C - I_B)R_5 = 0$$

One final simplification results whenever a branch contains a current source. In such a case, the current in that branch is known, and the branch can be ignored when the number of meshes is determined, so that the number of loop equations is reduced by one. Once the current in every branch is determined, all the voltages can be simply calculated from Ohm's law.

2.2   Superposition Theorem

Although the use of Kirchhoff's laws is the most general way to analyze circuits, the amount of effort required can usually be greatly reduced by making use of one of the circuit theorems, to be discussed in the next few pages. These theorems apply only to linear circuits, but they are nevertheless extremely useful. The first such theorem is called the **superposition theorem**, and it is useful whenever a linear circuit contains more than one source:

> *Superposition theorem:*   The current in a branch of a linear circuit is equal to the sum of the currents produced by each source, with the other sources set equal to zero.

The proof of the superposition theorem follows directly from the fact that Kirchhoff's laws applied to linear circuits always result in a set of linear equations, which can be reduced to a single linear equation in a single unknown. The unknown branch current can thus be written as a linear superposition of each of the source terms with an appropriate coefficient. In fact, the superposition theorem seems so reasonable that it is often mistakenly applied to nonlinear circuits.

The other trap in the application of the superposition theorem involves the meaning of setting a source equal to zero. One is often tempted to set a voltage source equal to zero by removing it entirely from the circuit. Recall, however, that setting a voltage source equal to zero means that the points in the circuit to which its terminals were connected must be kept at the same potential. The only way to do this is to replace the voltage source with a conductor (called a **short circuit**). A current source, on the other hand, is set equal to zero by leaving unconnected the points to which it was connected (called an **open circuit**). A short circuit causes the voltage to be zero; an open circuit causes the current to be zero.

These ideas can best be illustrated by means of an example. In figure 2.3(a) is a circuit containing a voltage and a current source. Although this circuit could easily be analyzed by circuit reduction or even by Kirchhoff's laws, we will use the superposition theorem to calculate the current I_2. First, in figure 2.3(b) we set the current source equal to zero by removing it. The current in I_2 due to the voltage source alone is just V divided by the equivalent resistance:

$$I_{21} = \frac{V}{R_1 + R_2}$$

This current is called the **partial current** in branch 2 due to source 1. In figure 2.3(c) the voltage source has been set equal to zero by short circuiting the

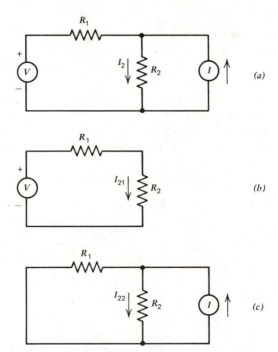

Fig. 2.3 The circuit in (a) can be analyzed using the supposition theorem by considering the simpler circuits in (b) and (c).

points to which it was connected. The resulting circuit is a current divider, and the resulting partial current given by equation 1.7 is

$$I_{22} = \frac{IR_1}{R_1 + R_2}$$

The superposition theorem then says that the total current is

$$I_2 = I_{21} + I_{22} = \frac{V + IR_1}{R_1 + R_2}$$

The current in R_1 could have been determined in a similar manner, with the result:

$$I_1 = \frac{V - IR_2}{R_1 + R_2}$$

2.3 Thevenin's Theorem

Perhaps the most useful of the circuit theorems is **Thevenin's theorem**:

> Any linear, two-terminal, dc network can be represented by a voltage source in series with a resistor.

A **network** is a group of circuit components (sources, resistors, etc.) connected together in some fashion. A **dc** (direct current) network is one in which the sources produce voltages and currents that are constant in time. A **two-terminal** network is one in which only two points in the circuit are available for observation and test. A **linear** network is one that contains only linear circuit components.

A way to visualize Thevenin's theorem is to imagine a black box (so as to conceal its contents) that contains an assortment of ideal sources and ideal resistors of arbitrary value connected in any complicated fashion. On the outside of the box are two terminals connected to any two points of the internal circuit. Such a network is depicted in figure 2.4(*a*). Thevenin's theorem says that no matter how complicated

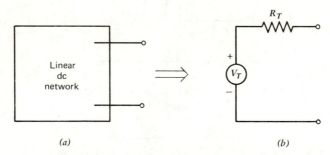

(*a*) (*b*)

Fig. 2.4 A linear, two-terminal, dc network (*a*) can be represented by a voltage source in series with a resistor (*b*).

the invisible network is, any measurements on the exposed terminals will be the same as that which would result if the network consisted of a single source and a single series resistor, as shown in figure 2.4(*b*). The value of the voltage and resistance will, of course, depend on what is the original network, but for a given network, there will be a unique value of V_T (called the **Thevenin equivalent voltage**) and R_T (called the **Thevenin equivalent resistance**).

The proof of Thevenin's theorem follows directly from the superposition theorem. If we connect a voltage source to the terminals of the network, as indicated in figure 2.5(*a*), and measure the current drawn by the source as the voltage is varied,

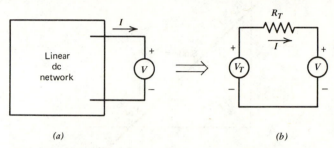

(*a*)   (*b*)

Fig. 2.5   The current I is the same function of V for circuit (*a*) and circuit (*b*) if V_T and R_T are chosen appropriately.

we know all that can be known about the circuit, short of opening the box and examining its contents. The superposition theorem says that the current I consists of two parts. If the external voltage V is set equal to zero, by short circuiting the terminals, the sources internal to the box will produce a partial current, called the **short circuit current**, I_{SC}. If the internal sources are set equal to zero, all that is left in the box is some combination of resistors that can be reduced to a single equivalent resistance R_T. The external source thus produces a partial current in the external branch given by $-V/R_T$. The negative sign arises because the arrow was drawn *into* the positive terminal of the voltage source, in contrast to the usual convention. The superposition theorem then gives

$$I = I_{SC} - \frac{V}{R_T} \tag{2.1}$$

A circuit that gives exactly this relation of V and I is the Thevenin equivalent circuit of figure 2.5(*b*):

$$I = \frac{V_T - V}{R_T} = \frac{V_T}{R_T} - \frac{V}{R_T}$$

provided V_T is adjusted so that

$$V_T = I_{SC} R_T \tag{2.2}$$

The voltage that appears across the terminals of the network when the current I is zero is called the **open circuit voltage** and is given by

$$V_{OC} = V_T = I_{SC}R_T \tag{2.3}$$

These quantities are indicated on the graph in figure 2.6. Note that a real voltage source [as in figure 1.4(*b*)] is just a Thevenin equivalent circuit.

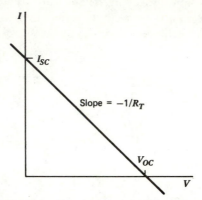

Fig. 2.6 Graph of I versus V for a linear, two-terminal, dc network.

A corollary of Thevenin's theorem is the following:

Any linear, two-terminal, passive, dc network can be represented by a single equivalent resistor.

By **passive**, we mean that the network contains no sources. In such a case, the open circuit voltage, and hence the Thevenin equivalent voltage, is necessarily zero. In such a case the equivalent resistance is just R_T.

Thevenin's theorem is most useful whenever the current in a particular resistor in a complicated linear network is to be calculated. We know that the circuit can be reduced to one in which the resistor whose current is to be calculated is connected to a Thevenin equivalent circuit. The resulting circuit is just a voltage divider, for which the solution is known. This fact, by itself, is of limited use, except to encourage us to apply circuit reduction techniques in an attempt to reduce the circuit to a single source and a single resistor.

The real usefulness of Thevenin's theorem comes from the fact that the Thevenin parameters V_T and R_T can be determined from the open circuit voltage and the short circuit current:

$$V_T = V_{OC} \tag{2.4}$$

$$R_T = \frac{V_{OC}}{I_{SC}} \tag{2.5}$$

Alternately, R_T can be determined by turning off all the sources and calculating the resistance between the network terminals.

As an example, we will use Thevenin's theorem to calculate the current in resistor R_5 in figure 2.1. First, the resistor R_5 is removed from the circuit, and the points B and C become the terminals of a linear, two-terminal, dc network. The two resistive branches of the resulting network are voltage dividers, and the open circuit voltage, and hence the Thevenin voltage, is

$$V_T = V_{OC} = V_{BD} - V_{CD} = \frac{VR_3}{R_1 + R_3} - \frac{VR_4}{R_2 + R_4} \tag{2.6}$$

where the node at D is used as a reference point.

The Thevenin resistance is calculated by setting $V = 0$, in which case the circuit reduces to the one shown in figure 2.7 which has an equivalent resistance of

$$R_T = \frac{R_1 R_3}{R_1 + R_3} + \frac{R_2 R_4}{R_2 + R_4} \tag{2.7}$$

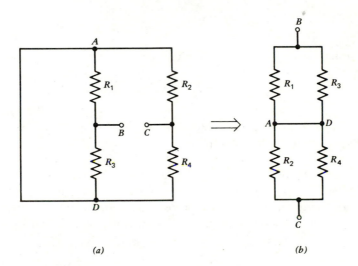

(a)   (b)

Fig. 2.7 The Thevenin equivalent resistance of the Wheatstone bridge can be determined by setting $V = 0$ (a) and redrawing the circuit as in (b).

The current in R_5 in figure 2.1 is then given by

$$I_5 = \frac{V_T}{R_T + R_5} \tag{2.8}$$

To derive this result from Kirchhoff's laws would have required several pages of algebra.

2.4 Norton's Theorem

A theorem closely related to Thevenin's theorem is **Norton's theorem**:

> Any linear, two-terminal, dc network can be represented by a current source in parallel with a resistor.

The proof of Norton's theorem has already been provided in section 1.5, where it was shown that the relation between V and I is the same for a voltage source with a series resistor and a current source with a parallel resistor. Norton's theorem thus follows directly from Thevenin's theorem.

Norton's theorem is used in the same way as Thevenin's theorem. The **Norton equivalent current** (I_N) for the network is obtained by short circuiting the terminals:

$$I_N = I_{SC} \tag{2.9}$$

and the **Norton equivalent resistance** (R_N) is obtained from

$$R_N = \frac{V_{OC}}{I_{SC}} \tag{2.10}$$

Note that in the two representations of a network, the following relations hold:

$$R_T = R_N \tag{2.11}$$

$$V_T = I_N R_N \tag{2.12}$$

Thevenin's theorem allows a circuit to be reduced to a voltage divider; Norton's theorem allows a circuit to be reduced to a current divider. Whenever one theorem is useful, the other is equally useful. The choice is largely one of taste.

As an example of the use of Norton's theorem, consider the circuit in figure 2.8(a). If we wish to calculate the current in R_2, we can remove R_2 and replace it with a short circuit. Since R_3 has no voltage across it and hence no current through it, the short circuit current and hence the Norton equivalent current is

$$I_N = I_{SC} = \frac{V}{R_1}$$

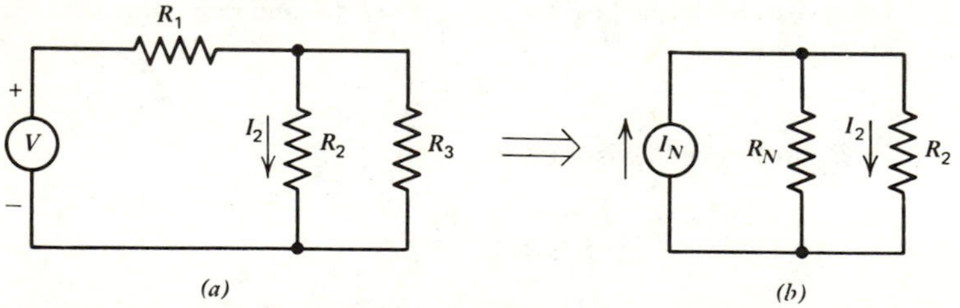

(a)   (b)

Fig. 2.8   The circuit in (a) can be reduced to a Norton equivalent circuit in (b) which is a current divider.

The Norton equivalent resistance is found by setting V equal to zero, which leaves R_1 and R_3 in parallel:

$$R_N = \frac{R_1 R_3}{R_1 + R_3}$$

Replacing the network with a Norton equivalent circuit and replacing R_2 give the circuit in figure 2.8(b), which is just a current divider. The current in I_2 is then given by the current divider relation:

$$I_2 = \frac{I_N R_N}{R_N + R_2} = \frac{V R_3}{R_1 R_2 + R_1 R_3 + R_2 R_3}$$

One might attempt to solve this problem also by circuit reduction in order to verify the above result and to compare the amount of work required.

2.5 Reciprocity Theorem

The final theorem that we will consider is called the **reciprocity theorem**:

> The partial current in branch 1 of a linear, dc circuit produced by a voltage source in branch 2 is the same as the partial current that would be produced in branch 2 by the same source if it were placed in branch 1.

The theorem is illustrated in figure 2.9, which shows a network with two pairs of

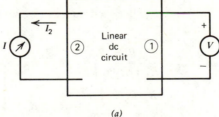

(b)

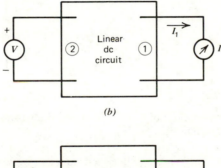

(a)

Fig. 2.9 The reciprocity theorem says that the partial current I_{12} in circuit (a) is the same as the partial current I_{21} in circuit (b).

terminals. A pair of terminals will be referred to as a **port**. Since the theorem involves only the partial currents produced by the external sources, a corollary of the reciprocity theorem is the following:

> In a linear, dc circuit with a single voltage source and an ammeter, the ammeter reading will remain the same if the ammeter and voltage source are interchanged.

It is assumed that both the voltage source and the ammeter are ideal. An alternate form of the reciprocity theorem is the same as the above but with "voltage source" replaced by "current source" and "ammeter" replaced by "voltmeter."

Rather than prove the reciprocity theorem in its most general form, we will prove it for a special case which will also serve to illustrate its usefulness. First, consider a linear, three-terminal, passive dc network. The three terminals can be paired off in three ways. Each pair of terminals must satisfy Thevenin's theorem and hence be representable as an equivalent resistance. Any circuit we can concoct that has the same three equivalent resistances as the actual circuit will be indistinguishable from it in terms of any external measurements we can make. The simplest representation for a three-terminal network must then contain three resistors, and there are only two ways these resistors can be connected. Figure 2.10 shows the so-called **Δ-connection**

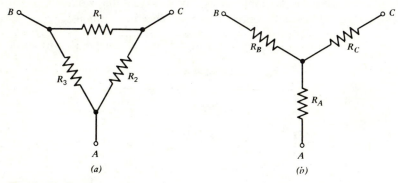

Fig. 2.10 Representations of a three-terminal, dc, passive network. (*a*) Δ-connection. (*b*) *Y*-connection.

and the **Y-connection**. The resistance between each combination of terminals can be calculated for the two circuits:

$$R_{AB} = \frac{R_3(R_1 + R_2)}{R_1 + R_2 + R_3} = R_A + R_B$$

$$R_{BC} = \frac{R_1(R_2 + R_3)}{R_1 + R_2 + R_3} = R_B + R_C$$

$$R_{CA} = \frac{R_2(R_1 + R_3)}{R_1 + R_2 + R_3} = R_C + R_A$$

$$(2.13)$$

With a bit of algebra, the **Δ-Y transformation** can be derived from the above:

$$R_A = \frac{R_2 R_3}{R_1 + R_2 + R_3}$$

$$R_B = \frac{R_1 R_3}{R_1 + R_2 + R_3}$$

$$R_C = \frac{R_1 R_2}{R_1 + R_2 + R_3}$$

$$\left. \begin{array}{l} \\ \\ \\ \end{array} \right\} \quad (2.14)$$

$$R_1 = \frac{R_A R_B + R_A R_C + R_B R_C}{R_A}$$

$$R_2 = \frac{R_A R_B + R_A R_C + R_B R_C}{R_B}$$

$$R_3 = \frac{R_A R_B + R_A R_C + R_B R_C}{R_C}$$

$$\left. \begin{array}{l} \\ \\ \\ \end{array} \right\} \quad (2.15)$$

The equivalence of the Δ- and Y-connection is sometimes a useful circuit reduction technique (see problem 2.12).

A three-terminal network is a special case of a two-port network in which two of the terminals are common. In the two-port representation, the Δ- and Y-connections are called the **π-network** and the **T-network**, respectively. These networks are shown in figure 2.11. Now suppose that such a passive, two-port network is connected to a pair of voltage sources and a pair of ammeters, as shown in figure 2.12. The

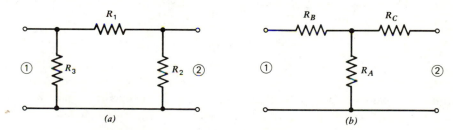

Fig. 2.11 Two-port network with a common terminal. (*a*) π-network. (*b*) T-network.

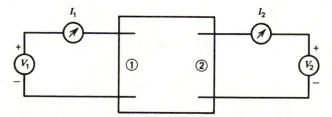

Fig. 2.12 Circuit for determining the R-parameters of a network with two ports.

superposition theorem says that the current in each branch is a superposition of the partial currents produced by each of the two sources:

$$
\left.
\begin{aligned}
I_1 = I_{11} + I_{12} = \frac{V_1}{R_{11}} + \frac{V_2}{R_{12}} \\
I_2 = I_{21} + I_{22} = \frac{V_1}{R_{21}} + \frac{V_2}{R_{22}}
\end{aligned}
\right\}
\qquad (2.16)
$$

The constants R_{11}, R_{12}, R_{21}, and R_{22} are called the **R-parameters** of the circuit. Inspection of equation 2.16 shows that the R-parameters are given by

$$
\frac{1}{R_{ij}} = \frac{\partial I_{ij}}{\partial V_j}
\qquad (2.17)
$$

The R-parameters are often written as a square matrix:

$$
\begin{bmatrix}
R_{11} & R_{12} \\
R_{21} & R_{22}
\end{bmatrix}
$$

Networks with n-ports can be described by an $n \times n$ matrix of R-parameters.

One use of the reciprocity theorem is to reduce the amount of work required to calculate the R-parameters for a circuit. Since a source at port 1 produces the same partial current at port 2 as the same source at port 2 would produce at port 1, the ratio I_{12}/V_2 is the same as the ratio I_{21}/V_1. This is equivalent to saying that $R_{12} = R_{21}$. More generally, for a multiport network, the matrix of R-parameters is diagonally symmetric:

$$
R_{ij} = R_{ji}
\qquad (2.18)
$$

As an example, we will calculate the R-parameters for the π-network in figure 2.11(a). In order to calculate R_{11} we set $V_2 = 0$ by short-circuiting the terminals in parallel with R_2. Since R_{11} is the ratio of the current at port 1 to the voltage at port 1 with port 2 shorted, it is just given by the parallel combination of R_1 and R_3:

$$
R_{11} = \frac{R_1 R_3}{R_1 + R_3}
$$

Similarly, R_{22} is determined by short-circuiting port 1 and calculating the resistance as seen by port 2:

$$
R_{22} = \frac{R_1 R_2}{R_1 + R_2}
$$

R_{21} is determined by placing a voltage source at 1 and an ammeter at 2. In such a circuit, neither R_2 nor R_3 affects the reading of the meter, and the ratio of current measured to voltage applied is

$$R_{21} = R_1$$

By the same argument, R_{12} is also equal to R_1, and so the reciprocity theorem ($R_{12} = R_{21}$) is satisfied.

2.6 Wheatstone Bridge

Before concluding the discussion of dc circuits, we will return to the Wheatstone bridge circuit mentioned earlier in the chapter. In figure 2.13, the Wheatstone bridge circuit of figure 2.1 is redrawn in a more customary manner, with resistor R_5 omitted.

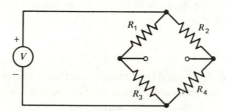

Fig. 2.13   Wheatstone bridge circuit.

Any network that can be drawn in such a diamond arrangement with a source connected at opposite nodes is called a **bridge circuit**. Bridge circuits are very useful in electronics. The Wheatstone bridge is one in which the bridge contains only resistors. The Wheatstone bridge is useful for making accurate resistance measurements, but it exemplifies a more general technique called the **null method** which is used throughout science and engineering for making highly accurate measurements.

Consider the general problem of measuring accurately the value of a resistor. The simplest method would be to place a known voltage across the resistor and with an ammeter measure the current drawn. But sources of accurately known voltage and meters of high accuracy (say, better than $\sim 1\%$) are quite expensive. On the other hand, resistors of 0.1% or better accuracy are easily manufactured, and they retain their accuracy indefinitely if not grossly abused. The Wheatstone bridge thus allows an unknown resistor to be compared with a standard resistor in such a way that its value can be determined to an accuracy that approaches that of the standard.

It has already been shown (section 2.3) that the circuit of figure 2.13 is equivalent to a Thevenin equivalent circuit with Thevenin parameters:

$$\left. \begin{aligned} V_T &= V\left(\frac{R_3}{R_1 + R_3} - \frac{R_4}{R_2 + R_4}\right) \\ R_T &= \frac{R_1 R_3}{R_1 + R_3} + \frac{R_2 R_4}{R_2 + R_4} \end{aligned} \right\} \tag{2.19}$$

The Thevenin equivalent voltage V_T is zero whenever

$$\frac{R_3}{R_1 + R_3} = \frac{R_4}{R_2 + R_4}$$

which can be rewritten as

$$R_2 R_3 = R_1 R_4 \tag{2.20}$$

Equation 2.20 is called the **balance condition** or **null condition**, and whenever it is satisfied the bridge is said to be balanced or nulled. The significance of the balance condition is that a meter placed across the output terminals of the bridge will read zero whenever equation 2.20 is satisfied, and it doesn't matter whether the meter is a voltmeter or an ammeter or anything between. Furthermore, the meter need not be accurate, but it should have high sensitivity. Similarly, the voltage source need be neither ideal nor accurately known, but its voltage should not be too low.

In practice, two of the resistors, say, R_1 and R_2, would be matched to high precision so that

$$R_1 = R_2$$

One of the resistors, say, R_3, would be a highly accurate variable resistor, and the last resistor, R_4, would be the unknown. The variable resistor would be adjusted until a galvanometer across the output of the bridge reads zero, at which point its value would just equal the value of the unknown. Since variable resistors of high accuracy are not common, a suitable substitute for R_3 would be a fixed resistor of high accuracy with a resistance slightly greater than required to balance the bridge. Another resistor of high value could then be added in parallel with it to achieve a balance. The value of this parallel resistor need not be known to such a high precision, since it contributes only slightly to the resistance of R_3.

Another use of a bridge circuit is to measure a small change in a quantity. Suppose we wish to determine how much the resistance of a resistor changes as a function of temperature. We could take a balanced Wheatstone bridge and keep all its resistors at a constant temperature except for one. If that one resistor, say, R_3, were heated up so that its resistance changed by an amount δR_3, a voltmeter at the output of the bridge would read a voltage

$$V_o = V \left(\frac{R_3 + \delta R_3}{R_1 + R_3 + \delta R_3} - \frac{R_4}{R_2 + R_4} \right)$$

If, for simplicity, we take all resistors to be the same,

$$R_1 = R_2 = R_3 = R_4$$

the output voltage is

$$V_o = V \left(\frac{R + \delta R}{2R + \delta R} - \frac{1}{2} \right)$$

$$= \frac{1}{2} V \left(\frac{1 + \delta R / R}{1 + \delta R / 2R} - 1 \right)$$

Using the useful relation,

$$\left(1 + \frac{\delta R}{R}\right)^n \simeq 1 + \frac{n\delta R}{R} \qquad (2.21)$$

for $|\delta R| \ll R$, the output voltage is seen to be proportional to δR:

$$V_o \simeq \frac{1}{4} V \frac{\delta R}{R} \qquad (2.22)$$

One could thus easily measure the fractional change in resistance $\delta R/R$ versus temperature. Most resistors have a very small variation of resistance with temperature ($<0.1\%/°C$), but resistors especially made to exhibit a high-temperature coefficient of resistance are called **thermistors**. A Wheatstone bridge consisting of three high-precision resistors and a thermistor could be used as a thermometer with an appropriately calibrated meter at its output.

2.7 Summary

In this chapter, techniques were introduced which simplify the analysis of circuits. In analyzing a circuit, one usually first does obvious circuit reduction such as combining resistors or sources that are in series or parallel. If the remaining circuit contains more than one source, the superposition theorem will probably be useful. Otherwise, either Thevenin's or Norton's theorem should probably be used. Multiterminal resistor networks are completely described by a matrix of R-parameters, and the reciprocity theorem simplifies the calculation of these parameters.

It is important to remember that the above theorems apply only to linear circuits. The last half of the book is filled with circuits for which these techniques fail miserably. For such circuits, Kirchhoff's laws may provide the only way to proceed. For linear circuits, the use of Kirchhoff's laws is seldom the easiest way to analyze a circuit. But for a computer or for a persistent human being who would rather manipulate a lot of algebra instead of a few circuit symbols, they should always provide an answer.

Problems

2.1 For the circuit below, write a set of independent current and voltage equations:

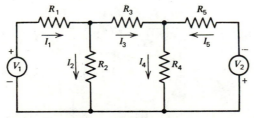

2.2   Use the loop current technique to calculate the current I_5 in figure 2.1 for the special case in which $R_5 = 0$.

2.3   For the circuit in figure 2.3(*a*), use Kirchhoff's laws to write an independent set of current and voltage equations, and solve these equations for the current I_2.

2.4   If the following equations are meshes of a certain circuit, can the circuit be reconstructed? If it can, then reconstruct it showing the direction of the currents and voltages. Write any other equations that are necessary for solving all the currents in the circuit.

$$V_1 = I_1 r_1 + I_1 R_1 + I_2 R_2$$

$$I_3 R_3 + I_4 R_4 + I_5 R_5 - I_2 R_2 = 0$$

$$I_3 R_3 + I_6 R_6 = 0$$

$$I_5 R_5 - I_7 R_7 = 0$$

$$V_2 = I_8 r_2 + I_8 R_8 - I_4 R_4$$

2.5   Use the superposition theorem to calculate the current I_3 in the circuit in problem 2.1, assuming $V_1 = 30$ V, $V_2 = 10$ V, $R_1 = R_2 = 100\ \Omega$, $R_3 = 50\ \Omega$, and $R_4 = R_5 = 200\ \Omega$.

2.6   A certain incandescent lamp has a nonlinear characteristic that can be approximated by $V = 280\ I^2$. Calculate the current in the lamp in the circuit below using Kirchhoff's voltage law, and compare your answer with the (incorrect) result of using the superposition theorem.

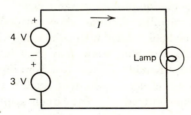

2.7   Calculate the Thevenin parameters of the circuit in problem 2.1 as seen by the resistor R_3 using the values given in problem 2.5. Use these values to calculate the current in I_3 for $R_3 = 50\ \Omega$.

2.8   Find the Thevenin equivalent of the circuit to the left of terminals AB below. What current will flow through terminals AB if they are shorted together?

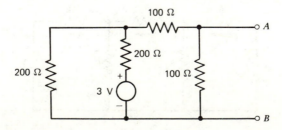

2.9   For the circuit below, show that the maximum power will be dissipated in R if $R = R_T$. This result is extremely useful in the design of circuits that must deliver the maximum possible power.

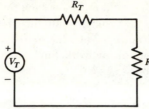

2.10   Determine the Norton equivalent current and the Norton equivalent resistance for the circuit in figure $2.3(a)$ as seen by the resistor R_2. Use Norton's theorem to determine the current I_2.

2.11   Find the Norton parameters for the circuit below. What is the output voltage V_{OC}?

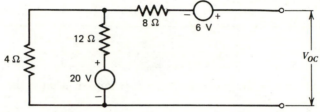

2.12   Find the equivalent resistance R of the circuit shown below:

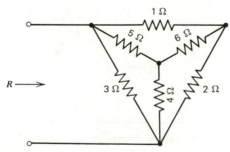

2.13   Calculate the R-parameters for the T-network in figure $2.11(b)$.

2.14   Calculate the R-parameters for the network below:

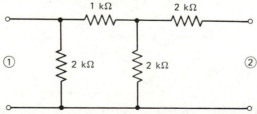

2.15   Construct a four-terminal network (two ports) consisting of three resistors connected in a π-network and having the following R-parameters:

$$\begin{bmatrix} 5\,\Omega & 10\,\Omega \\ 10\,\Omega & 8\,\Omega \end{bmatrix}$$

2.16   Calculate values of R_1, R_2, and R_3 below such that the voltage across the load resistor R_L would be one-tenth the value it would have if the network in the box were omitted and such that the resistance seen by the source and by the load are the same as they would have been without the network. Such a circuit is called a **T-pad**, and it is useful for voltage attenuation.

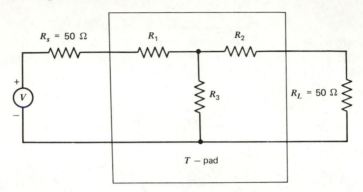

2.17   If the Wheatstone bridge in figure 2.1 is balanced, what is the resistance as seen by the source?

2.18   In the bridge circuit below, calculate the balance condition.

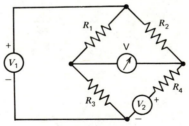

2.19   Draw the Thevenin equivalent circuit and calculate the values of V_T and R_T for the circuit below.

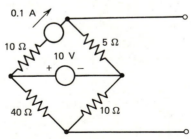

2.20   The **potentiometer circuit** shown below provides a null method for comparing an unknown voltage V with a known voltage V_S provided by a **standard**

cell. The voltage can be determined to high accuracy, without causing a current to flow in either the unknown source or the standard cell, by first adjusting for a null reading on the meter with the standard cell connected (α_1) and then readjusting it for a null with the unknown connected (α_2). Derive an expression for V, and show that the result is independent of V_0, r_1, r_2, and R.

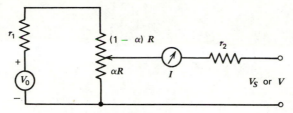

2.21 Shown in (a) below is a linear resistor network with three ports in which currents flow as indicated. With the network connected as shown in (b), calculate the current I_{SC}. With the network connected as shown in (c), calculate the voltage V_{OC}.

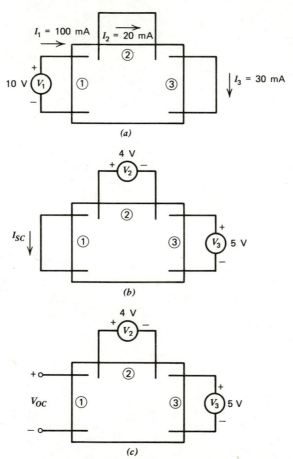

chapter 3

Transient Circuits

3.1 Time-Dependent Sources and Meters

In this chapter, we begin a study of circuits in which the currents and voltages vary in time. Most of the ideas encountered in the study of dc circuits remain valid for time-dependent circuits, but a number of new concepts will be encountered. Linear, dc circuits are described by linear, algebraic equations; whereas linear, time-dependent circuits are described by linear **differential** equations. Fortunately, for most of the cases of interest, these equations can be solved in a straightforward manner. In this chapter we will be concerned with a special case of time-dependent circuits in which the sources are dc but are turned on or off abruptly. Such circuits are called **transient** circuits, since the voltages and currents throughout the circuit readjust to a new dc value in a brief but nonnegligible time interval immediately following the change in state of the source. The initial condition is a dc circuit; the final condition is a different dc circuit; but the interval in between, while the circuit is readjusting to the new conditions, may exhibit complex behavior.

Perhaps the simplest way to turn a source on or off is by means of a **switch**. Figure 3.1(a) shows a voltage source connected to a resistor through a switch. The

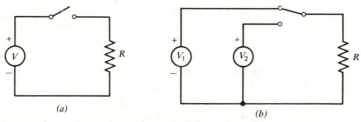

Fig. 3.1 Switches are useful for producing an abrupt change in a voltage or current. (a) Single-throw switch. (b) Double-throw switch.

switch is shown in its **open** position. If at some instant of time, say $t=0$, the switch is closed, a current will immediately begin to flow with a value given by Ohm's law:

$$I(t) = \begin{cases} 0 & t < 0 \\ V/R & t \geq 0 \end{cases}$$

This is an example of a **single-pole, single-throw switch**. Figure 3.1(b) shows an example of a **single-pole, double-throw switch** which at time $t = 0$ disconnects the resistor from V_1 and connects it to V_2. In practice, it is never possible to disconnect one circuit at precisely the same time another circuit is connected. To know for certain which will happen first, switches are made in two types. A **shorting switch** connects one circuit before disconnecting the other. A **nonshorting switch** disconnects one circuit before connecting the other. This difference is sometimes of great importance. For example, if the switch in figure 3.1(b) were a shorting type, an infinite current would flow through the switch briefly while the two sources are in parallel. In a circuit with real sources of low internal resistance, the large current could weld the switch contacts together and perhaps damage the sources as well. More complicated switches having multiple poles (controlling several circuits simultaneously) and multiple-throw (more than two positions) are quite common. A switch that can be remotely activated by energizing an electromagnet is called a **relay** or **contractor**.

Whereas the basic device for measuring dc voltages and currents is the D'Arsonval galvanometer, the basic device for measuring time-dependent voltages and currents is the **oscilloscope**. The heart of the oscilloscope is the **cathode ray tube** (CRT) shown schematically in figure 3.2. A filament (a) provides heat, which

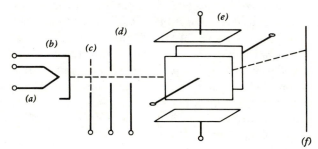

Fig. 3.2 Cathode ray tube. (a) Filament. (b) Cathode. (c) Control grid. (d) Focusing electrodes. (e) Deflection plates. (f) Fluorescent screen.

boils electrons off the cathode (b). A control grid (c) controls the intensity of the electron beam, which is focused by other electrodes (d). The beam can be deflected in either of two dimensions by deflection plates (e), and it finally strikes the fluorescent screen (f), causing it to emit light at a spot. The position of the spot can be moved up and down by varying the voltage on the vertical deflection plates or right and left by varying the voltage on the horizontal deflection plates. In this way the variation of one voltage as a function of another can be displayed graphically on the CRT screen. If the horizontal plates are connected to a voltage that increases linearly with time, a plot of the voltage at the vertical deflection plates as a function of time can be produced. Special **trigger** circuitry is usually provided to synchronize the horizontal sweep of the beam with the closing of a switch or with some feature of a repetitive waveform. The oscilloscope is thus basically a voltmeter with a high internal resistance (typically 1 MΩ), but it can also be used as an ammeter with an appropriate low-resistance shunt.

3.2 Capacitors

All the circuits encountered so far respond to time-varying sources in exactly the same way as they do to dc sources, namely, the currents and voltages everywhere in the circuit readjust instantly to any changes in the sources. All the equations written down so far are correct if we interpret all the variables as instantaneous quantities that may vary from one instant to the next. If that were the whole story, we would now be finished with the study of time-dependent circuits. However, two new circuit components enter the picture, and they greatly enhance the usefulness of electrical circuits. The first of these is called the **capacitor** (or **condenser** in older texts).

To understand the operation of a capacitor, imagine two large, parallel, conducting plates of area A separated a small distance d by an insulator which might be air but more typically is a dielectric such as paper, glass, plastic, oil, mica, or ceramic. Such a configuration is shown in figure 3.3. If a voltage V is applied between

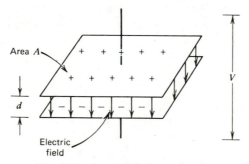

Fig. 3.3 A parallel plate capacitor.

the plates, an electric field $E = V/d$ will be produced. From Gauss's law, each plate must contain an equal and opposite electric charge given by

$$Q = \varepsilon A E = \frac{\varepsilon A V}{d} \tag{3.1}$$

where ε is the **permittivity** of the dielectric. Free space has a permittivity given by

$$\varepsilon_0 = 8.85 \times 10^{-12} \ \text{C}^2/\text{N} \cdot \text{m}^2$$

From the definition of current,

$$I = \frac{dQ}{dt} = \frac{\varepsilon A}{d} \frac{dV}{dt} \tag{3.2}$$

The constant $\varepsilon A/d$ is called the **capacitance**:

$$C = \frac{\varepsilon A}{d} \tag{3.3}$$

In the SI system, the unit of capacitance is the **farad** (abbreviated F), and is equal to 1 C/V:

$$1\text{ F} = 1\text{ C/V}$$

Typical capacitor values range from about 1 picofarad (pF, 10^{-12} farads) to about 1000 microfarads (1000 μF, 10^{-3} F).

Although the above derivation applies only to a particular configuration in which two large, parallel plates are separated by a small distance, any two conducting electrodes separated by an insulator will have a capacitance. The capacitance can be calculated exactly in only a few special cases such as the above. The capacitance is always the ratio of the charge on one of the electrodes to the voltage applied between the electrodes:

$$C = \frac{Q}{V} \tag{3.4}$$

A circuit element constructed in this way constitutes a capacitor, and from equation 3.2, we see that the relationship of the current through the capacitor to the voltage across its terminals is given by

$$I = C\frac{dV}{dt} \tag{3.5}$$

Although a real capacitor does not precisely obey the above equation, for a variety of reasons, we will define an ideal capacitor as one in which equation 3.5 holds exactly. Note the similarity to an ideal resistor in which Ohm's law is exactly satisfied. The symbols for an ideal capacitor are shown in figure 3.4. The quantity $C\,dV/dt$ has

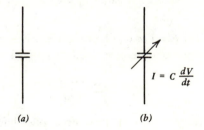

$$I = C\frac{dV}{dt}$$

(a)   (b)

Fig. 3.4 Symbols for ideal capacitors.
(a) Fixed. (b) Variable.

units of current and is called a **displacement current**, although it does not correspond to a flow of charge.

An ideal capacitor has several curious properties. First, note that if a constant voltage is placed across its terminals, equation 3.5 says that the current is zero. In a dc circuit, a capacitor thus behaves like an open circuit. On the other hand, if we try to change the voltage abruptly, the quantity dV/dt, and hence the current I, is infinite. Real circuits cannot have infinite currents, and so the voltage across a capacitor cannot change abruptly. In other words, for transients a capacitor behaves like a voltage source. A capacitor with zero voltage behaves transiently like a short circuit. Finally, energy cannot be dissipated in a capacitor. It can only be stored in the electric field, for later recovery. The energy stored in a capacitor is easily calculated:

$$W = \int VI\,dt = \int VC\frac{dV}{dt}\,dt = \int CV\,dV = \tfrac{1}{2}CV^2 \tag{3.6}$$

A capacitor with stored energy is said to be **charged**. A capacitor without stored energy ($V = 0$) is said to be **discharged**.

To conserve space, capacitors are usually made with numerous layers of conducting foil (usually aluminum) sandwiched between thin layers of insulation. Alternate layers of the foil are then connected together to provide the two terminals. The insulating material is carefully chosen according to its permittivity, breakdown voltage, and resistive power loss. The **relative permittivity**, $\varepsilon/\varepsilon_0$ is also called the **dielectric constant**. It varies typically from 2 for teflon to over 10^5 for some types of ceramic. The **breakdown voltage** for most dielectrics is several hundred volts per mil (1 mil = 0.001 inch). Capacitors are rated according to the voltage that can safely be applied across their terminals. Some dielectrics such as teflon and mica have extremely low power loss. Insulators with a large relative permittivity, such as some types of ceramic and most liquids, unfortunately have significant power loss.

The **electrolytic capacitor** is an especially compact design that uses aluminum or tantalum plates immersed in a semiliquid chemical compound which forms a thin, insulating, oxide layer on one of the electrodes. In addition to having a relatively high power loss, the electrolytic capacitor must be used in a circuit in which the sign of the voltage across its terminals is always the same (usually indicated by a + or − on the case of the capacitor). Such a capacitor is said to be **polarized**. Furthermore, the value of capacitance will vary considerably with voltage, temperature, age, and so on, for an electrolytic capacitor. But the cost per joule of energy storage capability is usually lower for an electrolytic capacitor than for any other type.

3.3 Inductors

A circuit element that behaves exactly opposite to the capacitor is the **inductor**, often called a **coil** or **choke**. To understand the operation of an inductor, imagine a circular coil of wire of area A with a constant number of turns per unit length (N/l) and a length long compared with its diameter. Such a coil, shown in figure 3.5, is called a **solenoid**. If a current I flows through the coil, Ampere's law allows us to

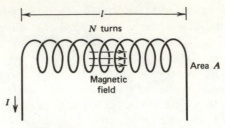

Fig. 3.5 A solenoidal inductor.

calculate the magnetic flux:

$$\Phi = BA = \frac{\mu NIA}{l} \tag{3.7}$$

where μ is the **permeability** of the material on which the coil is wound. For most materials (iron is a well-known exception) the permeability is close to the permeability of free space:

$$\mu_0 = 4\pi \times 10^{-7} \text{ N/A}^2$$

From Faraday's law, the voltage across the terminals of the coil is given by

$$V = N \frac{d\Phi}{dt} = \frac{\mu N^2 A}{l} \frac{dI}{dt} \tag{3.8}$$

The constant $\mu N^2 A/l$ is called the **inductance**:

$$L = \frac{\mu N^2 A}{l} \tag{3.9}$$

In the SI system, the unit of inductance is called the **henry** (abbreviated H) and is equal to one weber (a unit of magnetic flux; see Appendix D) per ampere:

$$1 \text{ H} = 1 \text{ W/A}$$

Typical inductor values range from about 1 μH (10^{-6} H) to about 1 H.

As with capacitance, the above derivation applies only to a particular coil configuration, but the concept of inductance is a very general one. All real components, including capacitors and resistors, have a certain inductance. Recall that a wire-wound resistor is very similar to the solenoid described above. The inductance is always the ratio of the magnetic flux linkage ($N\Phi$) to the current:

$$L = \frac{N\Phi}{I} \tag{3.10}$$

An ideal circuit component that contains only inductance is called an **inductor**, and from equation 3.8 we see that the relationship of voltage across the terminals of an inductor to the current through it is given by

$$V = L \frac{dI}{dt} \tag{3.11}$$

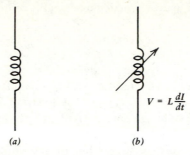

$$V = L\frac{dI}{dt}$$

(a)   *(b)*

Fig. 3.6 Symbols for ideal inductors. (*a*) Fixed. (*b*) Variable.

The symbols for an ideal inductor are shown in figure 3.6.

Like the capacitor, an ideal inductor has several curious properties. If a constant current passes through the inductor, equation 3.11 says that the voltage is zero. In a dc circuit, an inductor behaves like a short circuit. On the other hand, if V is to be finite, the current cannot change abruptly. For transients, the inductor thus behaves like a current source. Like a capacitor, an inductor cannot dissipate energy, but can only store it in the magnetic field. The energy stored in an inductor is

$$W = \int IV\,dt = \int IL\frac{dI}{dt}dt = \int LI\,dI = \tfrac{1}{2}LI^2 \qquad (3.12)$$

To conserve space, inductors are often wound on a toroidal iron core. Iron has a **relative permeability**, μ/μ_0, on the order of 1000. Unfortunately, an inductor with an iron core is far from ideal. To begin with, iron is an electrical conductor, and when a time varying current flows in the winding, a current is induced in the iron. This **eddy current** gives rise to resistive losses in addition to those of the wire used for the winding. Eddy currents can be reduced by laminating the iron and separating the laminations with an insulating varnish or shellac. Still better, the iron can be ground into a powder and mixed with an insulating binder. Some oxides of iron, nickel, and cobalt, called **ferrites**, also have a high relative permeability and a low electrical conductivity and thus have found widespread use in compact toroidal inductors.

A second difficulty with iron is that its permeability is not constant, but varies with the strength of the magnetic field and hence with the current in the windings. In fact, at sufficiently high magnetic fields, the core will saturate and its relative permeability will drop to a value near unity. Not only that, but the magnetic field in the iron depends on the past history of the current in the winding. This property of **remanence** is essential in a permanent magnet, but in an inductor it gives rise to additional losses, called **hysteresis losses**.

Variable inductors are usually made in the form of a short solenoid with a powdered iron or ferrite slug that can be screwed into or out of the form on which the coil is wound. Sometimes the slug is made of a conducting material such as brass, which has a relative permeability near unity, in which case eddy currents flow on the outside of the slug and eliminate magnetic flux from the center of the coil, reducing its effective area.

It is possible for the eddy current and hysteresis losses to be so large that the inductor behaves more like a resistor. Furthermore, there is always some capacitance between the turns of the inductor, and under some circumstances an inductor may act like a capacitor. This is a characteristic of all real circuit components. Whether a given component behaves more like a resistor, capacitor, or inductor depends on how it is made and how fast the voltages and currents are changing in time.

Both the ideal capacitor and the ideal inductor are, like the ideal resistor, linear components, since doubling the voltage doubles the current and vice versa. Note that we have now accumulated quite an assortment of conjugate pairs, as listed below:

voltage/current

series/parallel

loop/node

open/short

capacitance/inductance

charge/flux linkage

The existence of such pairs is a direct result of the symmetry of Maxwell's equations, which describe all electromagnetic phenomena.

3.4 Inductors and Capacitors in Combination

Just as circuit-reduction techniques are extremely useful with dc circuits, it is often possible to simplify circuits that contain more than one inductor or capacitor. Consider first the case of two inductors in series, as shown in figure 3.7(a). Since the

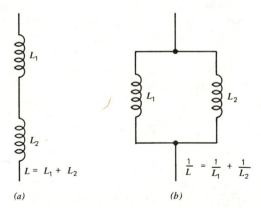

(a)   (b)

Fig. 3.7 Inductors in series (a) and parallel (b) add just as resistors do.

same current I must flow through both inductors, the total voltage drop is

$$V = L_1 \frac{dI}{dt} + L_2 \frac{dI}{dt} = (L_1 + L_2) \frac{dI}{dt}$$

Two inductors in series are therefore equivalent to a single inductor with an equivalent inductance given by

$$L = L_1 + L_2 \qquad (3.13)$$

Now consider the case of two inductors in parallel as shown in figure 3.7(b). Since the same voltage V appears across each, the total current is

$$I = \frac{1}{L_1} \int V \, dt + \frac{1}{L_2} \int V \, dt = \left(\frac{1}{L_1} + \frac{1}{L_2} \right) \int V \, dt$$

Two inductors in parallel are therefore equivalent to a single inductor with an equivalent inductance given by

$$\frac{1}{L} = \frac{1}{L_1} + \frac{1}{L_2} \qquad (3.14)$$

As with resistors, these relations can be generalized:

$$L = \sum_i L_i \qquad \text{(series)}$$

$$\frac{1}{L} = \sum_i \frac{1}{L_i} \qquad \text{(parallel)}$$

Now consider the case of two capacitors in parallel, as shown in figure 3.8(a). Since the same voltage V appears across each, the total current is

$$I = C_1 \frac{dV}{dt} + C_2 \frac{dV}{dt} = (C_1 + C_2) \frac{dV}{dt}$$

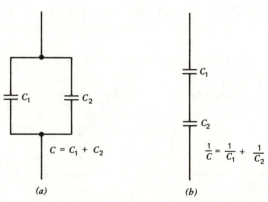

(a)

(b)

Fig. 3.8 Capacitors in parallel (a) and series (b) add opposite to the way resistors do.

Two capacitors in parallel are therefore equivalent to a single capacitor with an equivalent capacitance given by

$$C = C_1 + C_2 \tag{3.15}$$

Finally, consider the case of two capacitors in series as shown in figure 3.8(b). Since the same current I must flow through both capacitors, the total voltage drop is

$$V = \frac{1}{C_1} \int I \, dt + \frac{1}{C_2} \int I \, dt = \left(\frac{1}{C_1} + \frac{1}{C_2} \right) \int I \, dt$$

Two capacitors in series are therefore equivalent to a single capacitor with an equivalent capacitance given by

$$\frac{1}{C} = \frac{1}{C_1} + \frac{1}{C_2} \tag{3.16}$$

These relations can be generalized to give:

$$C = \sum_i C_i \quad \text{(parallel)}$$

$$\frac{1}{C} = \sum_i \frac{1}{C_i} \quad \text{(series)}$$

All these relations can easily be remembered simply by recalling that inductors combine the same way resistors do, but capacitors combine in the opposite (inverse) way. Similarly, one can form inductive and capacitive voltage and current dividers, provided the sources are time-dependent (see problem 3.4).

3.5 Series RC Circuit

We are now ready to consider circuits in which capacitors and inductors are combined with one another and with resistors. Since all these components are linear and since the relation of V and I for capacitors and inductors involve derivatives, the equations that result when Kirchhoff's laws are applied to such circuits are linear differential equations. The next few pages will review the techniques for solving such equations.

Consider first the transient circuit shown in figure 3.9(a) in which the source is

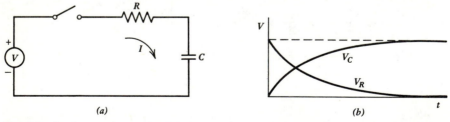

(a)   (b)

Fig. 3.9   In the transient series RC circuit in (a) in which the switch is closed at $t = 0$, the voltages adjust to a new equilibrium as indicated in (b).

dc, but at time $t = 0$ the switch is closed and remains closed until the circuit reaches a new equilibrium condition. Applying Kirchhoff's voltage law to the single loop that is formed after the switch is closed gives

$$V = IR + \frac{1}{C} \int I \, dt$$

As the first step in solving such an equation, we always eliminate any integrals by differentiating each term:

$$0 = R \frac{dI}{dt} + \frac{1}{C} I$$

where we have used the fact that V is a constant in this particular example. The next step always is to rewrite the equation in a standard form, in which all terms containing the unknown (I in this case) appear on the left of the equal sign with the highest derivative written first and without any multiplicative constants:

$$\frac{dI}{dt} + \frac{1}{RC} I = 0$$

This is an example of a **linear**, **first-order**, **homogeneous** differential equation. It is linear because the unknown appears only once to the first power in each term. It is first order because the highest derivative is the first, and it is homogeneous because the right-hand side, which would contain any terms not dependent on the unknown I, is zero.

The solution to all linear, first order, homogeneous differential equations is of the form

$$I = I_0 e^{\alpha t}$$

where the constant α is determined by substituting the solution back into the differential equation and solving the resulting algebraic equation:

$$\alpha + \frac{1}{RC} = 0$$

In this case the solution is

$$\alpha = -\frac{1}{RC}$$

The constant I_0 is determined from the initial condition at $t = 0$. The initial condition is easily determined from the fact that the voltage across a capacitor cannot change abruptly, and thus if the capacitor has zero voltage before the switch is closed, it will also have zero voltage immediately after the switch is closed. The capacitor initially behaves like a short circuit, and the initial current is

$$I(0) = I_0 = \frac{V}{R}$$

Therefore, the complete solution for the transient series RC circuit for an initially discharged capacitor is

$$I = \frac{V}{R} e^{-t/RC} \qquad (3.17)$$

The quantity RC is called the **time constant**, τ, for the circuit, since it has units of time and represents a characteristic time for the circuit to reach a new equilibrium condition after the switch is closed.

Once the current is known, the voltage across the resistor and capacitor can be easily determined:

$$V_R = IR = V e^{-t/RC}$$

$$V_C = \frac{1}{C} \int_0^t I\, dt = V(1 - e^{-t/RC})$$

A graph of these quantities is shown in figure 3.9(b). Whenever a result such as the above is obtained, it is always wise to check the limits $t = 0$ and $t = \infty$ to make sure that the result agrees with what one would expect for the appropriate dc circuits:

$$I(0) = \frac{V}{R} \qquad I(\infty) = 0$$

$$V_R(0) = V \qquad V_R(\infty) = 0$$

$$V_C(0) = 0 \qquad V_C(\infty) = V$$

Inspection of the circuit shows that these values are just what one would expect, since a capacitor initially $(t \ll RC)$ behaves like a short circuit, but after a long time $(t \gg RC)$ it behaves like an open circuit.

3.6 Series RL Circuit

The next example of a transient circuit is the series RL circuit shown in figure 3.10(a). As was the case for the series RC, the source is dc and the switch is closed at $t = 0$ and remains closed until the circuit readjusts to a new equilibrium.

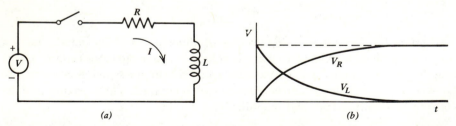

(a)   (b)

Fig. 3.10   In the transient series RL circuit in (a) in which the switch is closed at $t = 0$, the voltages adjust to a new equilibrium as indicated in (b).

Kirchhoff's voltage law gives for $t \geq 0$:

$$V = IR + L\frac{dI}{dt}$$

Rewriting in the standard form gives

$$\frac{dI}{dt} + \frac{R}{L}I = \frac{V}{L}$$

This is an example of a linear, first-order, **nonhomogeneous** differential equation, since the right-hand side is not zero. The solution to a linear, nonhomogeneous equation always consists of two parts:

$$I = I_h + I_p$$

The first part is called the **homogeneous solution**, and it is just the solution of the equation with the right-hand side set equal to zero:

$$\frac{dI_h}{dt} + \frac{R}{L}I_h = 0$$

We already know that such a linear, first-order, homogeneous equation has a solution

$$I_h = I_0 e^{\alpha t}$$

where α in this case is given by

$$\alpha = -\frac{R}{L}$$

If the term on the right-hand side of the nonhomogeneous equation is a constant independent of time, the **particular solution**, I_p, is also a constant, and its value can be easily determined by substituting into the original equation:

$$I_p = \frac{V}{R}$$

It is true that for whatever particular solution one finds to the equation, the homogeneous solution can always be added, since it gives zero when substituted into the left-hand side of the equation. It is usually needed, however, to satisfy the initial conditions. Putting the two parts of the solution together gives:

$$I = I_0 e^{-Rt/L} + \frac{V}{R}$$

The time constant τ for an RL circuit is given by $\tau = L/R$ in the same way that $\tau = RC$ for an RC circuit. All that remains is to calculate the constant I_0. To do that, we note that the current through an inductor cannot change abruptly, and the current just before the switch was closed was zero, and so

$$I(O) = I_0 + \frac{V}{R} = 0$$

which gives

$$I_0 = -\frac{V}{R}$$

The final solution for the current in a transient RL circuit is then

$$I = \frac{V}{R}(1 - e^{-Rt/L}) \qquad (3.18)$$

The voltage across the resistor and inductor are easily calculated:

$$V_R = IR = V(1 - e^{-Rt/L})$$

$$V_L = L\frac{dI}{dt} = Ve^{-Rt/L}$$

A graph of these quantities is shown in figure 3.10(*b*). The values of initial and final currents agree with what one would expect for the dc circuits in which the inductor is initially an open circuit but becomes a short circuit after a long time:

$$I(0) = 0 \qquad I(\infty) = \frac{V}{R}$$

$$V_R(0) = 0 \qquad V_R(\infty) = V$$

$$V_L(0) = V \qquad V_L(\infty) = 0$$

The reader should note the similarity of the RL and RC circuit behavior.

3.7 Series RLC Circuit

We now begin consideration of circuits that contain both a capacitor and an inductor. Such circuits are called **resonant circuits**. One of the simplest of such circuits is the series RLC circuit shown in figure 3.11. An even simpler circuit would result if the resistor were omitted, but there is always some resistance in a real series LC circuit, and so it would behave like the circuit of figure 3.10 in the limit of small resistance. We could also consider a series RLC circuit with a source, but that would only change the initial conditions. The general behavior of the circuit is the same with or without sources.

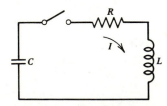

Fig. 3.11   Series *RLC* circuit.

Assume that the capacitor is charged to a voltage V_0, and then at $t = 0$ the switch is closed. Kirchhoff's voltage law gives for $t \geq 0$:

$$\frac{1}{C} \int I \, dt + IR + L \frac{dI}{dt} = 0$$

Rewriting in standard form gives

$$\frac{d^2 I}{dt^2} + \frac{R}{L} \frac{dI}{dt} + \frac{1}{LC} I = 0$$

This is an example of a linear, **second-order**, homogeneous differential equation. It is reasonable to guess that the solution is of the same form as for the first-order, homogeneous differential equation encountered earlier:

$$I = I_0 e^{\alpha t}$$

Substituting into the differential equation gives

$$\alpha^2 + \frac{R}{L} \alpha + \frac{1}{LC} = 0$$

Note that a solution of the form $e^{\alpha t}$ always reduces a linear, homogeneous differential equation to an algebraic equation in which first derivatives are replaced by α and second derivatives by α^2, and so forth. A linear, second-order, homogeneous, differential equation then becomes a quadratic algebraic equation, and so on. This particular algebraic equation has the following solutions:

$$\left. \begin{array}{l} \alpha_1 = -\dfrac{R}{2L} + \sqrt{\dfrac{R^2}{4L^2} - \dfrac{1}{LC}} \\[4mm] \alpha_2 = -\dfrac{R}{2L} - \sqrt{\dfrac{R^2}{4L^2} - \dfrac{1}{LC}} \end{array} \right\} \tag{3.19}$$

Since either value of α represents a solution to the original differential equation, the most general solution is one in which the two possible solutions are multiplied by arbitrary constants and added together:

$$I = I_1 e^{\alpha_1 t} + I_2 e^{\alpha_2 t}$$

The constants I_1 and I_2 must be determined from the initial conditions. An nth order differential equation will generally have n constants which must be determined from the initial conditions. In this case the constants can be evaluated from a knowledge of $I(0)$ and $dI/dt(0)$. Since the current in the inductor was zero for $t < 0$, and since it cannot change abruptly, we know that

$$I(0) = 0$$

Since the current is initially zero, the voltage across the resistor must be zero, and so the initial voltage across the inductor is the same as across the capacitor. Hence

$$\frac{dI}{dt}(O) = \frac{V_0}{L}$$

From these relations, we get

$$I_1 = -I_2 = \frac{V_0}{(\alpha_1 - \alpha_2)L}$$

The solution for the current in the series RLC circuit is thus

$$I = \frac{V_0}{(\alpha_1 - \alpha_2)L}(e^{\alpha_1 t} - e^{\alpha_2 t}) \tag{3.20}$$

where α_1 and α_2 are given by equation 3.19.

The solution in equation 3.20 has a quite different character, depending on whether the quantity under the square root in equation 3.19 is positive, zero, or negative. We will consider the three cases in turn:

Case 1: Overdamped

For $R^2 > 4L/C$ the quantity under the square root is positive, and both values of α are negative with $|\alpha_2| > |\alpha_1|$. The solution is the sum of a slowly decaying positive term and a more rapidly decaying negative term of equal initial magnitude. The solution is sketched in figure 3.12(a). An important limiting case is the one in which $R^2 \gg 4L/C$. In that limit the square root can be approximated as

$$\sqrt{\frac{R^2}{4L^2} - \frac{1}{LC}} = \frac{R}{2L}\sqrt{1 - \frac{4L}{R^2 C}} \simeq \frac{R}{2L} - \frac{1}{RC}$$

and the corresponding values of α are

$$\alpha_1 = -\frac{1}{RC} \qquad \text{and} \qquad \alpha_2 = -\frac{R}{L}$$

Then the current in equation 3.20 is

$$I \simeq \frac{V_0}{R}(e^{-t/RC} - e^{-Rt/L}) \tag{3.21}$$

In this limit the current rises very rapidly (in a time $\sim L/R$) to a value near V_0/R and then decays very slowly (in a time $\sim RC$) back to zero. Such a circuit closely resembles the RC circuit studied earlier (figure 3.9), as would be expected, since L was assumed small at the outset (compared to $R^2 C/4$).

Case 2: Critically Damped

For $R^2 = 4L/C$, the quantity under the square root is zero and $\alpha_1 = \alpha_2$. Equation 3.20 is then zero divided by zero, which is undefined. Therefore, the method of solution

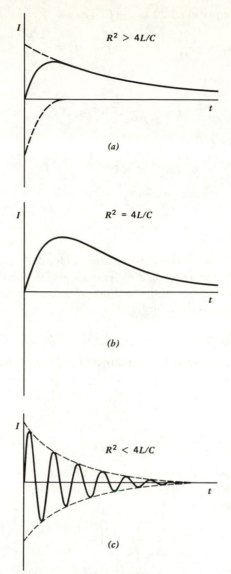

Fig. 3.12 Current versus time for a series *RLC* circuit. (*a*) Overdamped. (*b*) Critically damped. (*c*) Underdamped.

outlined above fails. A more productive approach is to let

$$\varepsilon = \sqrt{\frac{R^2}{4L^2} - \frac{1}{LC}}$$

and take the limit of equation 3.20 as $\varepsilon \to 0$. Then

$$\alpha_1 = -\frac{R}{2L} + \varepsilon \qquad \text{and} \qquad \alpha_2 = -\frac{R}{2L} - \varepsilon$$

and equation 3.20 becomes

$$I = \frac{V_0}{2\varepsilon L} e^{-\frac{Rt}{2L}} \left(e^{\varepsilon t} - e^{-\varepsilon t} \right)$$

Using the expansion

$$e^x \simeq 1 + x$$

for $|x| \ll 1$, the above equation becomes

$$I = \frac{V_0 t}{L} e^{-\frac{Rt}{2L}} \tag{3.22}$$

The same result could have been derived using l'Hôpital's rule (see Appendix E). This equation is sketched in figure 3.12(b). The shape of the curve is not very different from the overdamped case, except that it approaches zero as fast as possible without overshooting the t axis and going negative.

Case 3:   Underdamped

For $R^2 < 4L/C$, the quantity under the square root is negative, and α can be written as

$$\alpha = -\frac{R}{2L} \pm \frac{j}{\sqrt{LC}} \sqrt{1 - \frac{R^2 C}{4L}}$$

where

$$j = \sqrt{-1}$$

A j is used for the square root of -1 in electronics because the more usual symbol, i, is reserved for currents. Now it will be useful to define another symbol, ω, which we call the **angular frequency**:

$$\omega = \frac{1}{\sqrt{LC}} \sqrt{1 - \frac{R^2 C}{4L}} \tag{3.23}$$

Note that for $R^2 \ll 4L/C$, the angular frequency is

$$\omega \simeq \frac{1}{\sqrt{LC}}$$

and this approximation will usually suffice for most cases of interest. With these substitutions, equation 3.20 becomes

$$I = \frac{V_0}{2j\omega L} e^{-\frac{Rt}{2L}} \left(e^{j\omega t} - e^{-j\omega t} \right)$$

We now make use of the mathematical identity

$$e^{j\theta} = \cos \theta + j \sin \theta \qquad (3.24)$$

to express the current as follows:

$$I = \frac{V_0}{\omega L} e^{-Rt/2L} \sin \omega t \qquad (3.25)$$

This solution is of a very different form than the others, since it is oscillatory, with the oscillation amplitude decaying exponentially in time, as shown in figure 3.12(c). Although ω is referred to as the angular frequency, note that it has units of radians per second, and it is related to the usual frequency f which has units of cycles per second or **hertz** (abbreviated Hz) by

$$\omega = 2\pi f \qquad (3.26)$$

Similarly, the **period** of oscillation is

$$T = \frac{1}{f} = \frac{2\pi}{\omega} \qquad (3.27)$$

It is instructive to consider what happens to the energy in an underdamped, series RLC circuit. At $t = 0$, all the energy is stored in the capacitor. As the current increases, energy is dissipated in the resistor and stored in the inductor until one-quarter of a cycle has elapsed, at which time there is no energy left in the capacitor. But as time goes on, the energy in the inductor decreases, and the energy in the capacitor increases until one-half cycle has elapsed, at which time all the energy except that dissipated in the resistor is back in the capacitor. The energy continues to slosh back and forth, until it is eventually all dissipated by the resistor. The damping of an RLC circuit involves the conversion of ordered energy ($\frac{1}{2}CV^2$ and $\frac{1}{2}LI^2$) into disordered, thermal energy in the resistor, and so is just what would be expected from the second law of thermdynamics.

The **quality factor** of a resonant circuit is defined as the energy stored divided by the average energy dissipated per radian of oscillation:

$$Q = \frac{\omega W}{\bar{P}} \qquad (3.28)$$

where

$$\bar{P} = \frac{1}{T} \int_0^T I^2 R \, dt$$

It is left as an excercise (problem 3.12) to show that for a series RLC circuit the Q is given approximately by

$$Q \simeq \frac{\omega L}{R} \qquad (3.29)$$

Yet another equivalent definition of Q is the number of radians required for the stored energy to decay to $1/e$ of its original value. A series LC circuit without any resistance would have an infinite Q and would oscillate forever without damping. Real inductors always have some resistance, and circuits with Q greater than a few hundred are very difficult to construct.

The type of differential equation that describes the series RLC circuit is a very important one, because it appears with different variables in many areas of science and engineering. More generally, the system described by such an equation is called a **damped harmonic oscillator**. The shock absorbers on an automobile, for example, are part of a mechanical harmonic oscillator which is designed to be nearly critically damped. A thorough understanding of the series RLC circuit will provide considerable insight into a wide variety of such phenomena.

3.8 Summary

Transient circuits are circuits in which the sources are dc but are turned on or off abruptly. Transient circuits that contain only resistors behave in the same way as they would for dc. Two additional linear circuit components, the capacitor and the inductor, play important roles in transient circuits. The ideal inductor is defined by the relation

$$V = L\frac{dI}{dt}$$

and the ideal capacitor is defined by the relation

$$I = C\frac{dV}{dt}$$

in the same way that an ideal resistor is defined by Ohm's law,

$$V = IR$$

Inductors in series and parallel can be combined in the same way as resistors. Capacitors are combined in the opposite (inverse) way. Circuits that contain capacitors and inductors can be analyzed using Kirchhoff's laws, which lead to a set of simultaneous linear differential equations. For transient circuits, the solution consists of a homogeneous part that is proportional to $e^{\alpha t}$ and a particular part that is a constant. The differential equations can be reduced to algebraic equations from which the values of α can be determined. Constants will always appear in the solutions, and these will have to be determined from the initial conditions. The initial conditions are obtained from the circuit using the fact that the voltage across a capacitor and the current through an inductor cannot change abruptly.

Problems

3.1   Suppose two strips of conducting foil 1 meter long × 1 cm wide are alternated with strips of insulator 0.1 mm thick and relative permittivity of 10, and that the strips are rolled up into a cylinder with many layers. Calculate the capacitance.

3.2   Calculate the capacitance of two parallel plates each with an area of 100 cm² separated by a distance of 5 mm in air. What would the capacitance be if a 4-mm-thick conducting sheet were inserted between the plates?

3.3   Suppose an insulated wire with 1-mm diameter is close wound in a single layer on a 1-cm diameter × 10-cm-long iron core with a relative permeability of 1000. Calculate the inductance.

3.4   In the capacitive voltage divider below, the voltage V varies in time. Calculate the voltage V_1 across capacitor C_1.

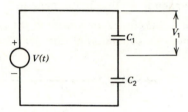

3.5   Calculate the current $I(t)$ and the voltage $V_C(t)$ across the capacitor in figure 3.9(a), assuming the capacitor has an initial voltage V_0.

3.6   In the circuit below, the switch is initially in position 1. At $t = 0$, the switch is moved to position 2. At $t = 1$ s, the switch is moved to position 3. Calculate $V_C(t)$ for $t \geq 0$, and sketch the result.

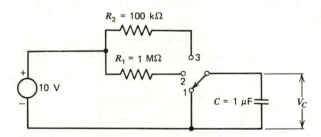

3.7   In the circuit below, the switch has been open for a long time, and then at $t = 0$ it is closed. Determine the current I and voltage V_3 just after the switch is closed ($t = 0$) and after a long time ($t \to \infty$).

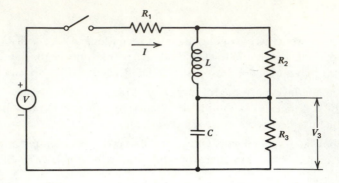

3.8   For the series RC circuit of figure 3.9, calculate the energy dissipated by the resistor and the energy stored in the capacitor as a function of time, and show that as $t \rightarrow \infty$, the energy stored is equal to the energy dissipated, independent of the values of R and C. Assume the capacitor is initially discharged.

3.9   After being open for a long time, the switch in the circuit below is closed at $t = 0$. Calculate the current I_L as a function of time for $t \geq 0$.

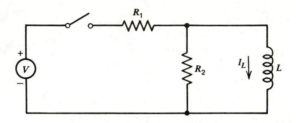

3.10   For the circuit in problem 3.9, assume the switch has been closed for a long time, and then at $t = 0$ it is opened. Calculate the voltage V_L across the inductor as a function of time for $t \geq 0$. If $V = 10$ V, $R_1 = 10 \, \Omega$, and $R_2 = 1 \, k\Omega$, what is the peak value of V_L?

3.11   Before the switch in the circuit below is closed, the capacitor C_1 is charged to a voltage $V_1(0)$, and C_2 is discharged, $V_2(0) = 0$. Calculate the final voltages $V_1(\infty)$ and $V_2(\infty)$.

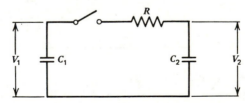

3.12   Show that for a series RLC circuit, the Q is given by $\omega L/R$ for $Q \gg 1$.

3.13   Determine the differential equation that describes the current I in the circuit below, and indicate the appropriate initial conditions if the switch is closed at $t = 0$.

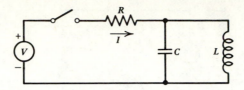

3.14 In the parallel RLC circuit below, the capacitor has an initial voltage V_0, and the switch is closed at $t=0$. Solve for the voltage V as a function of time if $R^2 \gg 4L/C$.

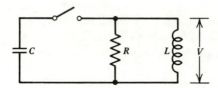

3.15 After being open for a long time, the switch in the circuit below is closed at $t=0$. Write a set of linearly independent equations that completely specify the behavior of the circuit, and combine these equations into a single differential equation with I_R as the only unknown.

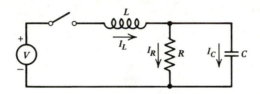

3.16 In the circuit below, the switch has been open for a long time and then is closed at $t=0$. Calculate I_L and V_C as a function of time for $t \geq 0$.

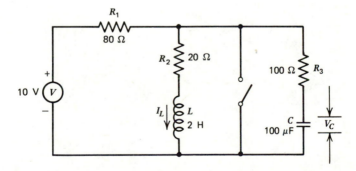

3.17 In the circuit below, the current increases linearly with time starting at $t=0$ such that $I=0$ for $t<0$ and $I=0.01\,t$ for $t \geq 0$. Find the voltage across the capacitor and the voltage across the inductor if at $t=0$ the capacitor is discharged.

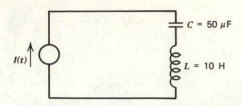

3.18 In the circuit below, the capacitor is initially charged to 1000 V, and both switches are open. At $t=0$, switch S_1 is closed. When the current in the inductor reaches its peak value, switch S_2 is closed. Sketch the voltage across the capacitor and the current through the inductor as a function of time, and show values of voltage, current, and time on your sketch. What would happen if S_2 were closed at a different time? Such a circuit is called a **crowbar**, and it is useful for producing intense, nearly constant, magnetic fields.

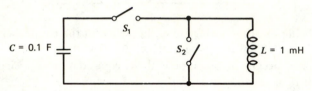

chapter 4

Sinusoidal
Circuits

4.1 Basic Definitions

In this chapter we will consider circuits in which the sources are sinusoidal functions of time. Such circuits are of particular importance because of the ease of producing sinusoidal time variations (as, for example, in the transient RLC circuits discussed in the preceding chapter), and because more complicated time variations can be treated as a superposition of sine waves (see the next chapter). The application of Kirchhoff's laws to such circuits will produce nonhomogeneous differential equations, but for linear circuits these equations can be transformed into complex, linear, algebraic equations.

Consider the circuit in figure 4.1(a) in which a sinusoidal voltage source,

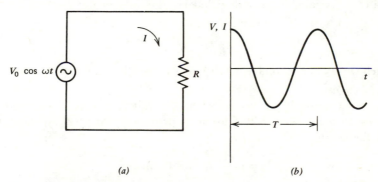

(a)   (b)

Fig. 4.1   A sinusoidal voltage source connected to a resistor (a) produces a current as in (b).

$V_0 \cos \omega t$, is connected to a resistor R. Whether we choose a cosine or a sine dependence for the voltage is arbitrary, since the shape of the waves are identical, and the only difference is in what we call $t = 0$. The cosine is more convenient, however, for what will follow. According to Ohm's law, the current in the circuit has a sinusoidal time dependence given by

$$I = \frac{V_0 \cos \omega t}{R}$$

The voltage and current are shown in figure 4.1(b). The **period** is the time required for the wave to repeat itself, and is given by

$$T = \frac{1}{f} = \frac{2\pi}{\omega} \tag{4.1}$$

The power dissipated by the resistor is a function of time. Of more interest is the *average* power. Since one cycle is representative of all others, we can average the power over a period to get

$$\bar{P} = \frac{1}{T} \int_0^T I^2 R \, dt$$

$$= \frac{1}{T} \int_0^T \frac{V_0^2 \cos^2 \omega t}{R} \, dt = \frac{V_0^2}{2R}$$

This result looks very similar to the usual definition of power in a dc circuit (equation 1.2) except for the factor of two. It is useful to define a **root mean square** (rms) voltage given by

$$V_{rms} = \left[\frac{1}{T} \int_0^T V^2(t) \, dt \right]^{1/2} \tag{4.2}$$

For a sinusoidal voltage, the rms value is given by $V_{rms} = V_0/\sqrt{2} \simeq 0.707 \, V_0$. The significance of the rms voltage is that if such a voltage is applied to a resistor, the same power will be dissipated as for a dc voltage of the same value:

$$\bar{P} = \frac{V_{rms}^2}{R} \tag{4.3}$$

Thus when we say that a voltage is 115 V ac, we usually mean that its value is given by $115 \sqrt{2} \cos \omega t$.

A more interesting case occurs when the voltage source is connected to a capacitor, as shown in figure 4.2(a). From the definition of an ideal capacitor, we can calculate the current:

$$I = C \frac{dV}{dt} = -\omega C V_0 \sin \omega t$$

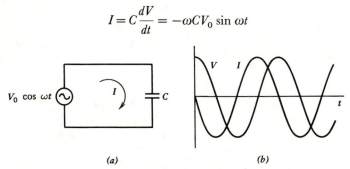

(a)   (b)

Fig. 4.2 A sinusoidal voltage source connected to a capacitor (a) produces a current as in (b).

The current has a maximum value (when $\sin \omega t = -1$) of

$$I_0 = \omega C V_0$$

but the maximum current does not occur at the same time as the maximum voltage. The voltage and current are sketched in figure 4.2(b). The ratio of peak voltage to peak current is like a resistance, since it has units of ohms, and it is called the **reactance**:

$$X = \frac{V_0}{I_0} \tag{4.4}$$

The reactance of a capacitor is thus

$$X_C = \frac{1}{\omega C} \tag{4.5}$$

The reciprocal of reactance is called **susceptance**, and, like conductance, is measured in units of siemens.

Note that the dc limit corresponds to setting $\omega = 0$, since $\cos(0) = 1$, and for such a case the capacitive reactance is infinite, and the capacitor behaves like an open circuit. On the other hand, at high frequencies ($\omega \to \infty$), the capacitive reactance is zero, and the capacitor behaves like a short circuit. Note also that the current can be expressed in terms of a cosine by

$$I = -I_0 \sin \omega t = I_0 \cos(\omega t + \phi)$$

where ϕ is called the **phase** of the current relative to the applied voltage. For the above case, the phase is 90° ($\pi/2$ rads). Note that phase, like voltage, is a relative quantity and that it can only be defined for two sinusoidal waves of the same frequency.

The energy that flows into the capacitor per unit time averaged over a cycle can be calculated in the same manner as for the resistor:

$$\bar{P} = \frac{1}{T} \int_0^T IV \, dt = -\frac{1}{T} \int_0^T \omega C V_0 \cos \omega t \sin \omega t \, dt$$

From the symmetry of the sine and cosine functions, we see that the energy that flows into the capacitor during the first and third quarter cycles is just balanced by the energy that flows out of the capacitor during the second and fourth quarter cycles, so that $\bar{P} = 0$. The capacitor neither dissipates nor permanently stores energy under such conditions, but just retains it temporarily and gives it back to the circuit a quarter cycle later.

Not surprisingly, an inductor behaves just the opposite of a capacitor. A circuit with a sinusoidal voltage source and an inductor is shown in figure 4.3(a). From the definition of an ideal inductor, we can calculate the current:

$$I = \frac{1}{L} \int V \, dt = \frac{V_0}{\omega L} \sin \omega t$$

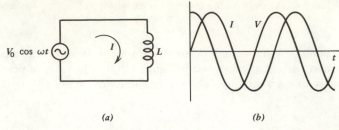

Fig. 4.3 A sinusoidal voltage source connected to an inductor (*a*) produces a current as in (*b*).

In performing an indefinite integral such as the above, there will, in general, be an arbitrary constant of integration that must be added to the result. In this case the constant would be the initial current in the inductor, which we take equal to zero. The voltage and current are sketched in figure 4.3(*b*). The current has a maximum value

$$I_0 = \frac{V_0}{\omega L}$$

and the reactance of an inductor is thus

$$X_L = \omega L \tag{4.6}$$

In the dc limit ($\omega = 0$), the inductive reactance is zero, and the inductor behaves like a short circuit. At high frequencies ($\omega \to \infty$), the inductive reactance is infinite, and the inductor behaves like an open circuit. The dc-limiting behavior of the capacitor and the inductor can easily be remembered by recalling their physical construction. As with the capacitor, the current in the inductor can be expressed in terms of a cosine by

$$I = I_0 \sin \omega t = I_0 \cos (\omega t + \phi)$$

but in this case the phase is $-90°$ ($-\pi/2$ rads). In a capacitor the current leads the voltage by 90°. In an inductor, the current lags the voltage by 90°. A useful way to remember this result, usually found in textbooks in which the symbol E is used for voltage instead of V, is with the phrase "ELI the ICE man," where L indicates an inductor and C a capacitor. As with the capacitor, the inductor does not dissipate power, but merely stores energy for release back to the circuit a quarter-cycle later.

4.2 Time-Domain Solutions

We are now ready to consider more challenging sinusoidal circuits such as the series RC circuit in figure 4.4(*a*). Applying Kirchhoff's voltage law to this circuit gives

$$V_0 \cos \omega t = IR + \frac{1}{C} \int I \, dt$$

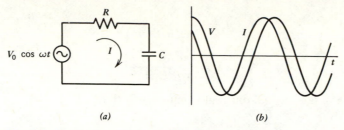

Fig. 4.4 In the sinusoidal series RC circuit in (a), the current leads the voltage by a phase $\phi = \tan^{-1}(1/\omega RC)$, as shown in (b).

As with transient circuits, we differentiate to eliminate any integrals and write the resulting equation in the standard form:

$$\frac{dI}{dt} + \frac{1}{RC}I = -\frac{\omega V_0}{R}\sin \omega t$$

This is a linear, first-order, nonhomogeneous differential equation similar to those encountered in the preceding chapter, except that the driving term on the right-hand side has a time dependence. In general, such an equation will have both an homogeneous and a particular solution. The homogeneous solution is needed to satisfy the initial condition when the source is first turned on. However, if we assume that the source has been on for a long time (much longer than $\tau = RC$ in this case), the transients, which decay exponentially, will have died away, and we need only be concerned with the particular solution. We might guess that the particular solution is either a sine or a cosine, but a quick inspection shows that neither of those, by itself, will satisfy the equation. A solution containing a bit of each is required:

$$I = I_1 \sin \omega t + I_2 \cos \omega t$$

Substituting the above into the differential equation gives

$$I_1\omega \cos \omega t - I_2\omega \sin \omega t + \frac{1}{RC}I_1 \sin \omega t + \frac{1}{RC}I_2 \cos \omega t = -\frac{\omega V_0}{R}\sin \omega t$$

The only way this equation can be satisfied for all values of t is if the coefficients of the sine and cosine terms separately add together:

$$\left.\begin{array}{c} I_1\omega + \dfrac{1}{RC}I_2 = 0 \\[3mm] -I_2\omega + \dfrac{1}{RC}I_1 = -\dfrac{\omega V_0}{R} \end{array}\right\}$$

A simple way to see this is to consider the cases $\omega t = 0$ and $\omega t = \pi/2$ for which the sine and cosine terms, respectively, vanish. Solving these two linear equations for I_1 and I_2 gives:

$$I_1 = -\frac{\omega C V_0}{\omega^2 R^2 C^2 + 1}$$

$$I_2 = \frac{\omega^2 R C^2 V_0}{\omega^2 R^2 C^2 + 1}$$

The solution of the original differential equation is thus

$$I = \frac{\omega C V_0}{\omega^2 R^2 C^2 + 1} (\omega R C \cos \omega t - \sin \omega t)$$

Note that in the limiting cases of $R = 0$ and $X_C = 0$, the above equation reduces to the results derived earlier for the circuits containing only a capacitor and only a resistor.

The current can also be written in terms of the cosine function alone by using the following trigonometric identity:

$$A \cos \omega t - B \sin \omega t = \sqrt{A^2 + B^2} \cos(\omega t + \phi) \tag{4.7}$$

where

$$\phi = \tan^{-1} \frac{B}{A}$$

The quantity $\tan^{-1}(B/A)$ is called the **inverse tangent** of B/A and is an angle whose tangent is B/A. The result is

$$I = \frac{\omega C V_0}{\sqrt{\omega^2 R^2 C^2 + 1}} \cos(\omega t + \phi)$$

where

$$\phi = \tan^{-1}\left(\frac{1}{\omega R C}\right)$$

A graph of the current and voltage for $\omega R C = 1$ is shown in figure 4.4(b). The voltage across the resistor and capacitor can now be determined:

$$V_R = IR = \frac{\omega R C V_0}{\sqrt{\omega^2 R^2 C^2 + 1}} \cos(\omega t + \phi)$$

$$V_C = \frac{1}{C} \int I \, dt = \frac{V_0}{\sqrt{\omega^2 R^2 C^2 + 1}} \sin(\omega t + \phi)$$

In performing the above indefinite integral for V_C, the constant of integration, which in this case corresponds to the initial voltage on the capacitor, has been taken equal to zero. If the capacitor had an initial voltage, it would decay to zero in a time $\tau = RC$. The neglect of the constant of integration in such a case is thus equivalent to the neglect of the homogeneous part of the solution, which is always justified after a sufficient time has lapsed and the circuit has reached a steady state.

The method of solution outlined above is called a **time-domain solution**, since the time dependences were carried throughout the calculation. Any linear differential equation with a sinusoidal driving term will have a solution that is just the sum of a term proportional to the sine and a term proportional to the cosine, and the coefficients of the terms can be determined as shown. However, for more complicated circuits, this method of solution can become very tedious. Fortunately, a shortcut exists, and that will be the subject of the next section.

4.3 Frequency-Domain Solutions

The example in the previous section illustrates an important property of linear circuits with a single sinusoidal source, namely, that the voltages and currents everywhere in the circuit are also sinusoidal with the same frequency as the source, but that the phase will vary throughout the circuit. Since linear circuits with several sinusoidal sources of different frequencies can be analyzed using the superposition theorem, the above principle is very useful. What it means is that we need not go to the trouble of calculating the time dependence of the unknown current or voltage, since we know that it will always be of the form $\cos(\omega t + \phi)$. All we need do is calculate the peak value and phase of the unknown. Such a method of solution is called a **frequency-domain solution**, since the equations will contain the angular frequency ω but not the time t.

A convenient method of analyzing circuits in the frequency domain makes use of the mathematics of complex numbers. It should be emphasized at the outset that currents and voltages are always real. When we write them with real and imaginary components, we are only introducing a mechanism for keeping track of the phase. The final answer must always be converted back to a form that does not contain any imaginary numbers.

Suppose we have a voltage source that produces a voltage $V_0 \cos \omega t$. We can represent this voltage as the real part of a vector of length V_0 at an angle ωt from the real axis in the complex plane, as shown in figure 4.5. The real part of such a vector will always be the length of the projection of that vector on the horizontal axis. The vector voltage is written as

$$V = V_0 e^{j\omega t}$$

and it rotates counterclockwise with angular frequency ω. All the other voltages and currents in the circuit containing such a source can also be similarly represented as vectors in the complex plane. Their length will correspond to their maximum value, their real part to their instantaneous value, and their angle with respect to the source vector will correspond to their phase. Since all the vectors rotate with the same frequency, it suffices to take a snapshot of the scene at any convenient time, since we know that at time t later, the whole scene will just be rotated through an angle ωt. Such a snapshot is called a **phasor diagram**, since the angles of the vectors represent phases. We usually choose to take the snapshot at time $t = 0$, when the source voltage $V_0 e^{j\omega t}$ lies along the real axis and the real part of V has its maximum value of V_0.

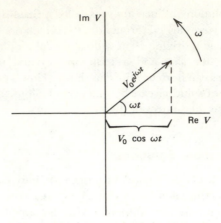

Fig. 4.5 A sinusoidal voltage V can be represented as a rotating vector in the complex plane.

If we apply a voltage $V_0 e^{j\omega t}$ to each of the three basic linear circuit components, we can calculate the current in each:

$$\text{Resistor:} \quad I = \frac{V}{R} = \frac{V_0}{R} e^{j\omega t}$$

$$\text{Capacitor:} \quad I = C\frac{dV}{dt} = j\omega C V_0 e^{j\omega t}$$

$$\text{Inductor:} \quad I = \frac{1}{L} \int V \, dt = \frac{V_0}{j\omega L} e^{j\omega t}$$

In each case the current has the same time dependence $(e^{j\omega t})$, but only in the case of the resistor does the phasor lie along the real axis. The capacitor current lies along the positive imaginary axis, and the inductor current lies along the negative imaginary axis (since $1/j = -j$). As the phasor diagram rotates in time, the current in the capacitor always leads the voltage by 90°, but the current in the inductor always lags the voltage by 90°.

The ratio of voltage to current in this representation is independent of time (the $e^{j\omega t}$ will cancel), but it will, in general, be a complex number, and it will be a function of frequency. This complex ratio is called the **impedance**, and it has units of ohms. For the three basic linear circuit components the impedance is given by:

$$\text{Resistor:} \quad Z = R$$

$$\text{Capacitor:} \quad Z = 1/j\omega C \qquad (4.8)$$

$$\text{Inductor:} \quad Z = j\omega L$$

For a resistor, the impedance is a real number equal to the resistance. For a capacitor or inductor, the impedance is an imaginary number with a magnitude equal to the reactance.

Capacitors and inductors, then, obey a relationship very similar to Ohm's law (equation 1.1), and it is sometimes called **ac Ohm's law**:

$$V = IZ \tag{4.9}$$

The ac Ohm's law reduces to the dc Ohm's law for circuits that contain only resistors, but for circuits with capacitors and inductors, the voltages and currents become complex numbers.

The reciprocal of impedance is called **admittance**. Like conductance and susceptance, admittance is measured in siemens. Note that admittance, like impedance, is a complex number, and that the angle that the admittance vector makes with the real axis is equal and opposite to the angle that the corresponding impedance vector makes with the real axis (see problem 4.2).

The usefulness of the impedance concept is that all of the circuit-reduction techniques and circuit theorems for dc circuits can be applied to linear sinusoidal ac circuits if the impedances of the various components are substituted into the equations as if all the components were resistors. One need never solve a differential equation for steady-state, linear, sinusoidal circuits. The equations will be complex algebraic equations, and the solution will be a complex number corresponding to a vector in the complex plane. The length of the vector will be the peak value of the quantity, and the angle that it makes with the real axis will be the phase.

As an example, consider the circuit in figure 4.6, in which a sinusoidal voltage,

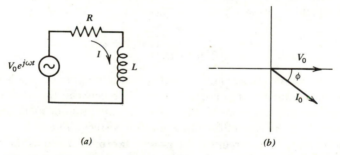

(a)   (b)

Fig. 4.6   The circuit in (a) has a phasor diagram as in (b) where ϕ $= \tan(-\omega L/R)$.

represented by $V_0 e^{j\omega t}$, is applied to a series RL circuit. Treating the resistor and inductor like two resistors and using ac Ohm's law, we can immediately write the phasor current:

$$I_0 = \frac{V_0}{R + j\omega L}$$

Whenever a complex expression has a j in its denominator, we always multiply and divide by the complex conjugate of the denominator. (The complex conjugate is the same expression with j replaced by $-j$):

$$I_0 = \frac{V_0}{R+j\omega L}\left(\frac{R-j\omega L}{R-j\omega L}\right) = \frac{(R-j\omega L)V_0}{R^2+\omega^2 L^2}$$

This trick will always reduce the expression to one of the form $A+jB$, which can be written as

$$A+jB = \sqrt{A^2+B^2}\,e^{j\phi} \tag{4.10}$$

where

$$\phi = \tan^{-1}\frac{B}{A}.$$

A is the real part of the complex number, and B is the imaginary part. For the above case, the phasor current is

$$I_0 = \frac{V_0}{\sqrt{R^2+\omega^2 L^2}}\,e^{i\phi}$$

where

$$\phi = \tan^{-1}\frac{-\omega L}{R}$$

Transformation back to the time domain, after the result has been expressed in the form of equation 4.10, is always accomplished by simply replacing $e^{j\phi}$ with $\cos(\omega t + \phi)$, which in this case gives

$$I = \frac{V_0}{\sqrt{R^2+\omega^2 L^2}}\cos(\omega t + \phi)$$

A phasor diagram of the voltage and current is shown in figure 4.6(b). Note that the current lags the voltage by an amount intermediate between the case of a resistor alone ($0°$) and an inductor alone ($-90°$). The consideration of such limiting cases will help ensure that the sign of the phase has the correct value.

A useful quantity in ac circuits is the **power factor**, defined as the ratio of the power dissipated by an impedance to the **apparent power** that would result from multiplying the rms voltage by the rms current. It is given by the cosine of the phase angle between the voltage and the current or by the ratio of the real part to the magnitude of the impedance:

$$\text{Power factor} = \frac{\text{Re}(Z)}{|Z|} = \cos\phi$$

The magnitude of a complex number is the square root of the sum of the squares of its real and imaginary parts:

$$|Z| = \sqrt{(\text{Re}Z)^2 + (\text{Im}Z)^2}$$

It can also be determined by multiplying the number by its complex conjugate and taking the square root.

The power factor is a fraction less than unity and is often expressed as a percentage. In power distribution systems a power factor near 100% is desired to provide the consumer with the maximum useful power while minimizing the ohmic losses in the transmission lines.

4.4 Series RLC Circuit

As another example of a frequency-domain solution, we will analyze the important case of a series RLC circuit connected to a sinusoidal voltage source, as shown in figure 4.7(a). The phasor current is just the source voltage divided by the total impedance:

$$I_0 = \frac{V_0}{R + j\omega L + 1/j\omega C}$$

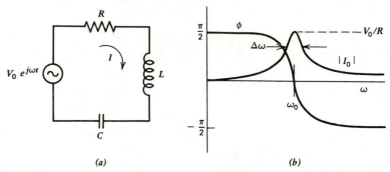

(a)   *(b)*

Fig. 4.7 The series RLC circuit in (a) exhibits resonant behavior as shown in (b).

Multiplying and dividing by the complex conjugate of the denominator gives

$$I_0 = \frac{[R - j(\omega L - 1/\omega C)]V_0}{R^2 + (\omega L - 1/\omega C)^2} = \frac{V_0}{\sqrt{R^2 + (\omega L - 1/\omega C)^2}} e^{j\phi}$$

where

$$\phi = \tan^{-1}\left(\frac{1/\omega C - \omega L}{R}\right)$$

In the time domain, the current is

$$I = \frac{V_0}{\sqrt{R^2 + (\omega L - 1/\omega C)^2}} \cos(\omega t + \phi)$$

Note that when R is small, the magnitude of the phasor current $|I_0|$ is large whenever ω is equal to the angular resonant frequency,

$$\omega_0 = \frac{1}{\sqrt{LC}} \qquad (4.11)$$

For $\omega = \omega_0$, the phase ϕ is zero, the current I_0 is V_0/R, and the circuit looks purely resistive. What has happened is that the impedances of the inductor ($j\omega L$) and capacitor ($1/j\omega C$) exactly cancel, and their series combination acts like a short circuit. Below resonance ($\omega < \omega_0$) the circuit looks capacitive, and above resonance ($\omega > \omega_0$) the circuit looks inductive. The magnitude and phase of the current are shown in figure 4.7(b). The smaller the resistance becomes, the narrower and higher becomes the curve of the current in figure 4.7(b). In fact, the width, $\Delta\omega$, of the curve at the points where the current is $1/\sqrt{2}$ ($\sim 70\%$) of its peak value (called the **half-power points**, since $P = I^2 R$) is another measure of the Q of the circuit. It will be left as a problem (4.5) to show that for $Q \gg 1$, the Q is given by

$$Q \simeq \frac{\omega_0}{\Delta\omega} \qquad (4.12)$$

Although the capacitor and inductor combination behaves like a short circuit at resonance, this does not mean that no voltage appears across them individually. In fact, from the value of the current in the circuit, we can easily calculate the voltage across all three components *at resonance*:

$$V_R = IR = V_0 \cos \omega_0 t$$

$$V_L = L \frac{dI}{dt} = -\frac{\omega_0 L}{R} V_0 \sin \omega_0 t$$

$$V_C = \frac{1}{C} \int I \, dt = \frac{\omega_0 L}{R} V_0 \sin \omega_0 t$$

The voltage across the resistor is the same as the source voltage, but the inductor and capacitor have equal and opposite voltages 90° out of phase with the source and larger than the source voltage by a factor $\omega_0 L/R$. Hence another interpretation of Q is the ratio of the voltage across one of the reactive components to the voltage across the resistance in a resonant, sinusoidal, series, RLC circuit. The fact that a sinusoidal voltage can be greatly magnified by such a simple circuit often comes as a shocking revelation!

One should note the relative algebraic simplicity of the sinusoidal series RLC circuit as compared to the transient series RLC circuit described in section 3.7. This comparison illustrates the great usefulness of the impedance concept and serves as an apt reward for one who is not frightened by the use of complex numbers. Note, however, that impedance is a purely sinusoidal concept, and so it should not be applied to the transient circuits of the previous chapter.

4.5 Filter Circuits

Linear circuit components can be used to construct circuits that pass certain frequencies while rejecting others. Such circuits are called **filters**, and their uses are numerous. We will consider here several common examples of filter circuits.

Consider first the series RL circuit in figure 4.8(a) in which a voltage $V_{in} = V_0 e^{j\omega t}$

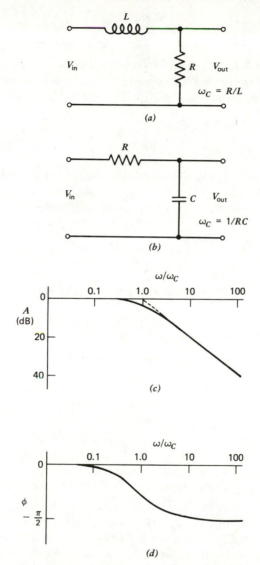

Fig. 4.8 The low-pass filters in (a) and (b) produce an attenuation (c) and phase (d) that vary with frequency.

is applied at the input. The output voltage can be calculated using the voltage divider relation:

$$V_{out} = \frac{RV_{in}}{R + j\omega L} = \frac{R^2 - j\omega RL}{R^2 + \omega^2 L^2} V_{in}$$

The ratio of the magnitudes of the two voltages is

$$\left| \frac{V_{out}}{V_{in}} \right| = \frac{R}{\sqrt{R^2 + \omega^2 L^2}} = \frac{1}{\sqrt{1 + (\omega L/R)^2}}$$

For ω small ($\ll R/L$), the input voltage appears at the output unattenuated, ($V_{out} = V_{in}$), but for ω large ($\gg R/L$), very little output voltage appears. Such a circuit is called a **low-pass filter**, and the quality R/L is called the **angular cutoff frequency**,

$$\omega_c = \frac{R}{L} \tag{4.13}$$

since it is the angular frequency at which the output voltage drops to $1/\sqrt{2}$ (half power) of the input value.

The series RC circuit in figure 4.8(b) can be analyzed in the same way with the result:

$$\left| \frac{V_{out}}{V_{in}} \right| = \frac{1}{\sqrt{1 + (\omega RC)^2}}$$

This circuit behaves exactly the same, except the angular cutoff frequency is

$$\omega_c = \frac{1}{RC} \tag{4.14}$$

The ratio $|V_{out}/V_{in}|$ is called the **attenuation** and is often expressed in dimensionless units called **decibels** (abbreviated dB):

$$A_{dB} = -20 \log_{10} \left| \frac{V_{out}}{V_{in}} \right| \tag{4.15}$$

An attenuation of 10 dB thus means that the power delivered to the load (which is proportional to V^2) is reduced by a factor of 10. An attenuation of 20 dB would correspond to a power reduction of $10^2 = 100$, and so on. A graph of A versus the normalized angular frequency ω/ω_c for the circuits described above is shown in figure 4.8(c). The point at which $\omega = \omega_c$ is called the **3-dB point**, since $A = 20 \log_{10} \sqrt{2} \simeq 3$ dB. At high frequencies, A increases by ~ 6 dB/octave or 20 dB/decade. An **octave** is a musical term meaning a factor of 2 in frequency. A **decade** is a factor of 10. A change in sound level of 1 dB is about the smallest change that can be detected by the human ear. Note that decibels add, so that if a circuit with 10 dB of attenuation is followed by a circuit with 20 dB of attenuation, the total attenuation is 30 dB, provided the second circuit does not alter the attenuation of the first.

Another interesting quantity is the phase of the output relative to the input for the two circuits. For both cases the phase is

$$\phi = \tan^{-1}\frac{-\omega}{\omega_C} \tag{4.16}$$

For $\omega \ll \omega_C$, the phase shift is negligible, but for $\omega \gg \omega_C$, it approaches $-90°$. At $\omega = \omega_C$, the phase shift is $-45°$. The phase as a function of ω/ω_C is plotted in figure 4.8(d).

The opposite behavior is produced by the circuit in figure 4.9(a), for which the

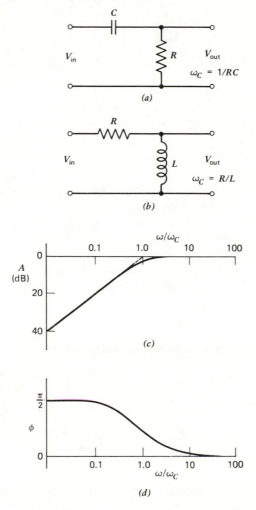

Fig. 4.9 The high-pass filters in (a) and (b) produce an attenuation (c) and phase (d) that vary with frequency.

output voltage is

$$V_{out} = \frac{RV_{in}}{R + 1/j\omega C} = \frac{\omega^2 R^2 C^2 + j\omega RC}{1 + \omega^2 R^2 C^2} V_{in}$$

The ratio of output to input voltage is

$$\left| \frac{V_{out}}{V_{in}} \right| = \frac{\omega RC}{\sqrt{1 + \omega^2 R^2 C^2}}$$

For $\omega_c = 1/RC$, the above expression becomes

$$\left| \frac{V_{out}}{V_{in}} \right| = \frac{1}{\sqrt{1 + \omega_c^2/\omega^2}}$$

The circuit in figure 4.9(b) gives the same result provided

$$\omega_c = \frac{R}{L}$$

In these cases the phase shift is

$$\phi = \tan^{-1} \frac{\omega_c}{\omega} \tag{4.17}$$

which is opposite to the low-pass filter. These circuits are called **high-pass filters**, and their attenuation and phase are plotted in figure 4.9(c) and (d).

More complicated filter circuits can be constructed which have almost any desired attenuation and phase characteristics, although a phase shift inevitably occurs whenever the attenuation varies with frequency. Two common examples are the **resonant filter** (problem 4.10) and the **notch filter** (problem 4.11). The art of filter design is highly developed, and digital computers are often used to optimize the design of filters for special applications.

4.6 Integrators, Differentiators, and Attenuators

The simple low- and high-pass filter circuits in the previous section can be used to produce an output voltage that approximates the integral or derivative of the input voltage. For example, applying Kirchhoff's current law to the RC circuit in figure 4.8(b) gives

$$\frac{V_{in} - V_{out}}{R} = C \frac{dV_{out}}{dt}$$

If $V_{out} \ll V_{in}$, the term on the left is approximately V_{in}/R, and the above expression can be integrated to give:

$$V_{out} \simeq \frac{1}{RC} \int V_{in}\, dt \tag{4.18}$$

Such a circuit is called an **RC integrator**. The circuit in figure 4.8(a) also produces an output proportional to the integral of the input (provided $V_{out} \ll V_{in}$) given by

$$V_{out} \simeq \frac{R}{L} \int V_{in} \, dt \tag{4.19}$$

RC integrators are more common than RL integrators, because capacitors are usually cheaper, smaller, and more nearly ideal than inductors. In both cases it is important that V_{out} be kept small, and this is achieved by making the time constant (RC or L/R, respectively) very long compared to the period or duration of the signal that is to be integrated.

In a similar manner the circuits in figure 4.9 can be used to produce an output voltage that approximates the derivative of the input voltage. Applying Kirchhoff's current law to the RC circuit in figure 4.9(a) gives

$$C \frac{d(V_{in} - V_{out})}{dt} = \frac{V_{out}}{R}$$

If $V_{out} \ll V_{in}$, the above expression becomes

$$V_{out} \simeq RC \frac{dV_{in}}{dt} \tag{4.20}$$

Such a circuit is called an **RC differentiator**. Similarly, the RL circuit in figure 4.9(b) also produces an output proportional to the derivative of the input (provided $V_{out} \ll V_{in}$) given by

$$V_{out} \simeq \frac{L}{R} \frac{dV_{in}}{dt} \tag{4.21}$$

V_{out} is kept small compared with V_{in} by making the time constant (RC or L/R) very short compared to the period or duration of the signal that is to be differentiated. It is important to realize that the integrator and differentiator work for any time dependent waveform and not just for sine waves. One should verify, however, that the low-and high-pass filters of the previous section do integrate and differentiate sine waves in the appropriate limit.

Often it is desirable to attenuate a sinusoidal voltage by an amount that is independent of frequency. It will be shown in the next chapter that this is equivalent to reducing the size of a nonsinusoidal voltage without distorting its shape. In theory, one could simply use a resistive voltage divider, since its output voltage is independent of frequency. In practice, there is always some stray capacitance in a real circuit, and eventually a frequency is reached at which the voltage divider behaves like either a low- or a high-pass filter. This difficulty can be overcome by using the circuit in figure 4.10 which is called a **compensated attenuator**. At low frequencies the circuit behaves like an ordinary resistive voltage divider, but at high frequencies the capacitive reactance dominates, and the circuit behaves like a capacitive voltage divider. It is left as an exercise (problem 4.19) to show that the attenuation is independent of frequency, provided

$$R_1 C_1 = R_2 C_2 \tag{4.22}$$

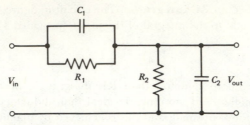

Fig. 4.10 In the compensated attenuator, the attenuation is independent of frequency provided $R_1 C_1 = R_2 C_2$.

In practice, one of the capacitors is usually variable, so that the attenuator can be adjusted to compensate for any stray capacitance.

Such compensated attenuators are often used at the input of an oscilloscope to raise the input resistance and lower the input capacitance so as to make the oscilloscope into a more nearly ideal voltmeter. A necessary penalty, however, is a decrease in sensitivity of the oscilloscope to input voltage. Such tradeoffs of two desirable quantities are commonly encountered in electronic circuit design.

4.7 Transformers

The list of circuit components considered so far is relatively short: sources, meters, resistors, capacitors, and inductors. In this section we introduce a new linear circuit component called the **transformer**. It differs from all the others in that it is a four-terminal rather than a two-terminal device. A transformer is nothing more than two inductors placed close enough together that some of the magnetic flux of one inductor links the other.

Imagine two inductors wound on the same laminated iron core, as shown in figure 4.11. Iron is used to increase the inductance of the windings and to ensure that

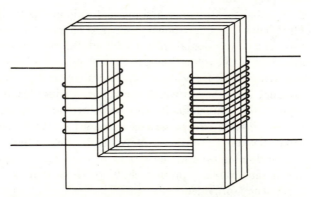

Fig. 4.11 A transformer can be made by winding two inductors on the same iron core.

most of the magnetic flux is shared by both windings. The iron is laminated to reduce the eddy currents that would otherwise flow in the conducting iron. Eddy-current losses increase with the square of the frequency and with the square of the thickness of the laminations. By contrast, hysteresis losses (see section 3.3) are proportional to frequency. Transformers are normally designed so that the ohmic losses in the windings and the core losses are comparable at the highest frequency that is to be used. At high frequencies ($\gtrsim 100$ kHz), a ferrite or air core would normally be used. Usually, transformers are made with the windings directly on top of one another rather than as shown in figure 4.11 to ensure good coupling between the windings.

If we arbitrarily designate one of the windings as the **primary** and connect it to an ac voltage source, V_{in}, a magnetic flux is produced in the iron core:

$$\Phi = \int \frac{V_{in}}{N_p} \, dt$$

where N_p is the number of turns on the primary. But according to Faraday's law, this flux produces a voltage in the other winding (called the **secondary**) given by

$$V_{out} = N_s \frac{d\Phi}{dt}$$

where N_s is the number of turns on the secondary. Combining the above two equations gives

$$\frac{V_{out}}{V_{in}} = \frac{N_s}{N_p} \tag{4.23}$$

A transformer thus has the property of producing an output voltage proportional to the input voltage with a proportionality constant that is independent of frequency and equal to the turns ratio.

If the secondary is open circuited, the primary current is given by

$$I_M = \frac{V_{in}}{j\omega L_p} \tag{4.24}$$

where L_p is the primary inductance. This current is called the **magnetizing current**, and it is usually small in a properly designed transformer. Since I_M always becomes large at very low frequencies ($\omega \to 0$), a transformer is inherently an ac device. It is useful to define an ideal transformer as one in which equation 4.23 holds exactly and in which the primary inductance is sufficiently large that the magnetizing current is negligibly small for the frequencies of interest. The symbols for an ideal transformer are shown in figure 4.12.

One use for a transformer is for impedance matching. Imagine that the secondary of an ideal transformer is connected to a resistor R_L and the primary to a sinusoidal voltage source with an rms value V_p. The rms voltage across R_L will be

$$V_s = \frac{N_s V_p}{N_p}$$

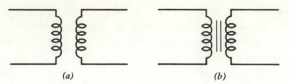

Fig. 4.12 Symbols for ideal transformer. (*a*) Air core. (*b*) Iron core.

and the power dissipated by the resistor is

$$P = \frac{V_s^2}{R_L} = \frac{N_s^2 V_p^2}{N_p^2 R_L}$$

Since an ideal transformer cannot dissipate power (for the same reason that an inductor cannot), the same power must be supplied by the source, so that the rms current in the source, and hence in the primary, is

$$I_p = \frac{P}{V_p} = \frac{N_s^2 V_p}{N_p^2 R_L}$$

The source therefore thinks it is connected to a resistor with a value

$$R = \frac{V_p}{I_p} = \left(\frac{N_p}{N_s}\right)^2 R_L \tag{4.25}$$

The same result holds for an arbitrary impedance Z_L at the secondary:

$$Z = \left(\frac{N_p}{N_s}\right)^2 Z_L \tag{4.26}$$

Matching the source impedance to the load impedance is important as a means of transferring the maximum power to the load (see problem 2.9). When the load is partly reactive, the maximum power is delivered when the source impedance is equal to the complex conjugate of the load impedance (see problem 4.21). In such a case the source and load reactances cancel, and the current in the load is maximum.

Real transformers depart from this ideal behavior in a number of ways. In addition to the finite inductance, the windings also have resistance and capacitance, and the coupling between windings is never perfect. A more realistic representation of a transformer in terms of ideal components is shown in figure 4.13, in which C_p, R_p and C_s, R_s represent the capacitance and resistance of the primary and secondary windings, respectively. The resistor R_c represents the core losses. Unlike an ordinary resistor, its value is dependent on the frequency. The quantity k is called the **coupling coefficient** and varies from zero for two isolated inductors to one for an ideal transformer. It is just the fraction of the magnetic flux produced by the primary that links the secondary. A well-designed iron core transformer might have $k \sim 95\%$. The quantity $(1 - k^2 L_p)$ is called the **leakage inductance**.

The construction of a real transformer always involves a compromise. One would

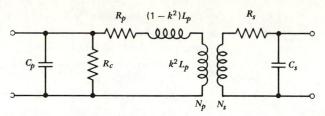

Fig. 4.13 Representation of real transformer in terms of ideal components.

like a large primary inductance to reduce the magnetizing current, but then the leakage inductance becomes large, since the coupling coefficient is always somewhat less than one. Transformers can be made that are reasonably ideal over two or three decades of frequency, which makes them barely suitable for use in high-fidelity audio equipment. Not shown in figure 4.13, but often of importance, is the capacitance between the primary and secondary windings. Again, a compromise is required, because a transformer constructed to have a small leakage inductance will generally have a large interwinding capacitance. Sometimes transformers are designed with an interwinding conducting shield that can be grounded to prevent capacitive coupling between the primary and secondary. Such a shield will, however, enhance the capacitance between each winding and ground.

The ability of a transformer to convert ac voltages from one level to another with negligible ($\lesssim$ a few %) loss of power illustrates one of the reasons why ac circuits are normally preferred over dc circuits in power distribution systems. Since the resistive power losses in the lines that run from the power plant to the consumer increase with the square of the rms current, it is a distinct advantage to operate such systems at high voltages and low currents. Transformers at the power plant increase the voltage to values in excess of 100 kV, and transformers reduce the voltage at the other end to values that are safer and more convenient. Alternating currents are also easier to produce using rotating machines (generators). Although ac voltages are more convenient for many applications such as synchronous motors (as used in electric clocks, turntables, and tape drives), it is often necessary to convert the ac to a dc voltage. Circuits for performing this function are described in Chapter 6.

4.8 Summary

Linear circuits which contain sources that vary sinusoidally in time are described by linear differential equations that have solutions of the form $\cos(\omega t + \phi)$, where ϕ is the phase. The simplest way to analyze such circuits is to transform into the frequency domain where all voltages and currents are represented by stationary vectors, called phasors, in the complex plane. The length and direction of a phasor specify the magnitude and phase of the quantity that it represents. In the frequency domain an inductor is represented by an impedance $j\omega L$, and a capacitor is represented by an impedance $1/j\omega C$. The rules for combining impedances for ac circuits are the same as the rules for combining resistances in dc circuits.

These techniques were used to analyze series RC, RL, and RLC circuits. The RC and RL circuits are useful as filters and as integrators and differentiators. The RLC circuit can be used as a resonant filter. Compensated attenuators can be made which have an attenuation that is independent of frequency. A new linear circuit element, the transformer, was introduced. It is useful for changing the magnitude of an impedance.

The analysis of sinusoidal ac circuits is, in principle, no more difficult than dc circuits, except that one calculates with two-dimensional vectors (called phasors) rather than with scalars. Kirchhoff's current law for ac circuits says that the sum of the vector currents flowing into a node is zero. Kirchhoff's voltage law for ac circuits says that the sum of the vector voltage drops around a loop is zero. But a vector is just a set of scalars that represent its components. The use of complex numbers is a convenient way to express the components of a two-dimensional vector. Although the mathematical expressions are often long and unwieldy, the algebra involved is quite straightforward.

Problems

4.1   Calculate the current $I(t)$ for $t \geq 0$ in the circuit in figure 4.4, assuming the source is turned on abruptly at $t = 0$ with the capacitor initially discharged.

4.2   Suppose that the impedance of a circuit is given by $Z = A + jB$. Show that the admittance Y is given by $|Y| = 1/|Z|$ and that the angles that Y and Z make with the real axis are equal and opposite.

4.3   Calculate the current for the circuit in figure 4.4, using a frequency-domain solution, assuming the source has been on for a long time.

4.4   Calculate the impedance of the circuit below. At what angular frequency is the circuit purely resistive?

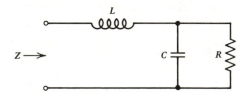

4.5   Show that equation 4.12 is consistent with an earlier definition of $Q = \omega L/R$ for a series RLC circuit provided $Q \gg 1$.

4.6   Calculate the peak value of the voltage across the inductor in figure 4.7(a), assuming $V_0 = 10$ V, $\omega = 2\pi \times 10^3$ s^{-1}, $R = 1\ \Omega$, $L = 25$ mH, and $C = 1\ \mu$F.

4.7   Calculate the phase of the voltage across the inductor relative to the source in figure 4.7(a) for the values given in problem 4.6.

4.8   Determine the resistances R_L and R_C such that the impedance Z in the circuit below is real for all frequencies. Determine the phase between the driving voltage and the current through R_L at a frequency $f = (1000/2\pi)$ Hz.

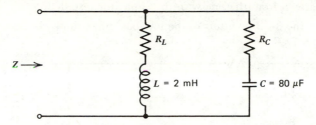

4.9   For the circuit below calculate the Thevenin equivalent voltage and the Thevenin equivalent impedance. Show how the Thevenin equivalent circuit could be constructed using individual circuit elements (resistors, inductors, etc.) in series, and indicate the required values.

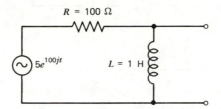

4.10   Calculate and sketch the ratio $|V_{out}/V_{in}|$ and the phase ϕ of the output relative to the input as a function of angular frequency for the resonant filter shown below:

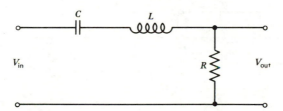

4.11   Calculate and sketch the ratio $|V_{out}/V_{in}|$ and the phase ϕ of the output relative to the input as a function of angular frequency for the notch filter shown below:

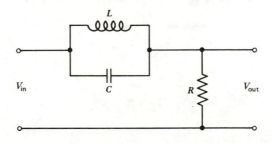

4.12 An electric motor has a power factor of 80% and draws an rms current of 10 A when connected to a 120-V, 60-Hz power line. What value of capacitor should be placed in parallel with the motor to minimize the current drawn from the line? What rms current is drawn from the line with the capacitor installed?

4.13 Calculate the 3-dB point ω_C of the low-pass filter shown below:

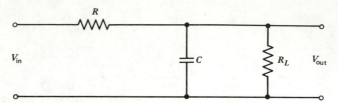

4.14 Calculate the 3-dB point ω_C and the number of dB per decade attenuation for $\omega \gg \omega_C$ for the filter below:

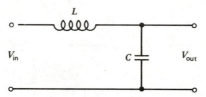

4.15 The input circuit of an oscilloscope often has a switch as shown below that allows only the ac component of a voltage to be observed. Calculate the lower and upper 3-dB points if the oscilloscope is ac coupled to a source with a 1000-Ω internal resistance.

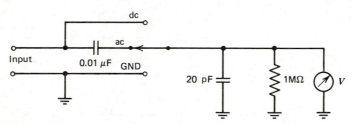

4.16 The circuit below is called an **all-pass filter** or **phase shifter**. Calculate $|V_{out}/V_{in}|$ and the phase ϕ of the output relative to the input as a function of angular frequency. What value does ϕ have for $\omega = 0$, $1/RC$, and ∞?

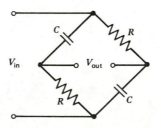

4.17 The circuit below is called a **Wien bridge**, and it is useful for measuring small changes in frequency. Calculate the balance conditions.

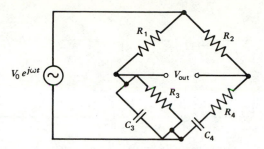

4.18 The circuit below is called a **twin-tee**. It is useful because it exhibits resonant behavior without the use of an inductor. Calculate the frequency f at which the current I is zero. What is the phase of I relative to V_0 for a frequency just below the resonance?

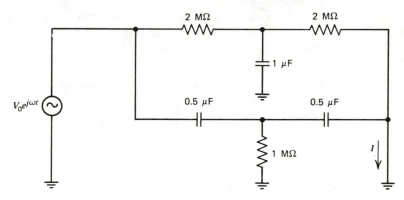

4.19 Show that if $R_1 C_1 = R_2 C_2$ in the circuit in figure 4.10, the attenuation is independent of frequency and is given by the usual voltage divider relation.

4.20 In the circuit below an ideal transformer is used to connect the output of a hi-fi amplifier to a speaker. The amplifier can be considered as a Thevenin equivalent circuit with a 200-Ω source resistance and the speaker can be considered as an 8-Ω resistive load. What turns ratio will result in maximum power delivered to the speaker? If the amplifier delivers 8 W to the speaker, what is the rms current in the primary for the above calculated turns ratio?

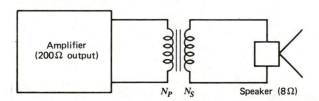

4.21 Calculate the value of L for which the current I is in phase with the source voltage for the ideal transformer shown below. For this value of L calculate the rms value of the current I.

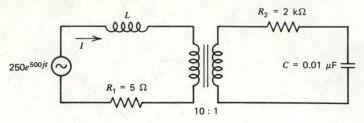

4.22 Show that the coupling coefficient of an otherwise ideal transformer can be determined by connecting the primary to an ac voltage source and measuring the ratio of the primary current with the secondary open circuited to the primary current with the secondary short circuited.

4.23 In the circuit below estimate the values of L_p and k required such that the 3-dB points of V will occur at 20 and 20,000 Hz. (Hint: In the low-frequency limit the leakage inductance can be ignored. In the high-frequency limit the magnetizing current can be ignored.)

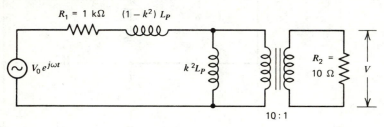

chapter 5

Nonsinusoidal and Distributed Circuits

5.1 Fourier Series

In this chapter we will consider linear circuits in which the sources are time dependent but not sinusoidal and circuits in which the circuit elements are not discrete components but where the inductance, capacitance, and resistance are distributed in a continuous manner. A time-dependent voltage or current is either **periodic** or **nonperiodic**. Figure 5.1 shows an example of a periodic waveform with period T. The wave is assumed to continue indefinitely in both the $+t$ and $-t$ directions. A periodic function can be displaced by one period, and the resulting function is identical to the original function:

$$V(t \pm T) = V(t)$$

A periodic waveform can be represented as a **Fourier series** of sines and cosines:

$$V(t) = \frac{a_0}{2} + \sum_{n=1}^{\infty} (a_n \cos n\omega_0 t + b_n \sin n\omega_0 t) \tag{5.1}$$

where ω_0 is called the **fundamental angular frequency**,

$$\omega_0 = \frac{2\pi}{T} \tag{5.2}$$

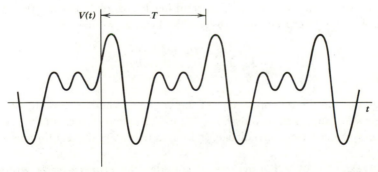

Fig. 5.1 Example of a periodic voltage with period T.

$2\omega_0$ is called the **second harmonic**, and so on. The constants a_n and b_n are determined from

$$a_n = \frac{2}{T} \int_{-T/2}^{T/2} V(t) \cos n\omega_0 t \, dt \tag{5.3}$$

$$b_n = \frac{2}{T} \int_{-T/2}^{T/2} V(t) \sin n\omega_0 t \, dt \tag{5.4}$$

The constant term $a_0/2$ is the average value of $V(t)$. The superposition theorem then allows us to analyze any linear circuit having periodic sources by considering the behavior of the circuit for each of the sinusoidal components of the Fourier series. Although most of the examples that we will use have voltage or current as the dependent variable and time as the independent variable, the Fourier methods are very general and apply to any sufficiently smooth function, $f(t)$.

For the same reason that it was useful to describe sinusoidal voltages and currents as complex numbers, it is useful to express a general periodic waveform as a sum of complex numbers:

$$V(t) = \sum_{n=-\infty}^{\infty} C_n e^{jn\omega_0 t} \tag{5.5}$$

This representation is equivalent to equation 5.1, as can be seen by substituting $e^{j\theta} = \cos\theta + j\sin\theta$ into equation 5.5 (see problem 5.5). By allowing both positive and negative frequencies ($n > 0$ and $n < 0$), it is possible to choose the C_n in such a way that the summation is always a real number. The value of C_n can be determined by multiplying both sides of equation 5.5 by $e^{-jm\omega_0 t}$, where m is an integer, and then integrating over a period. Only the term with $m = n$ survives, and the result is

$$C_n = \frac{1}{T} \int_{-T/2}^{T/2} V(t) e^{-jn\omega_0 t} \, dt \tag{5.6}$$

Note that C_{-n} is the complex conjugate of C_n, and so the imaginary parts of equation 5.5 will always cancel, and the resulting $V(t)$ is real. The $n = 0$ term has a particularly simple interpretation. It is just the average value of $V(t)$:

$$C_0 = \frac{1}{T} \int_{-T/2}^{T/2} V(t) \, dt \tag{5.7}$$

and corresponds to the dc component of the voltage. Whether the integrals in the above expressions are over the interval $-T/2$ to $T/2$ or some other interval such as 0 to T is purely a matter of convenience, so long as the interval is continuous and has duration T.

As an example of a Fourier series, consider the **square-wave voltage** in figure 5.2. The constants C_n can be determined from equation 5.6 by breaking the

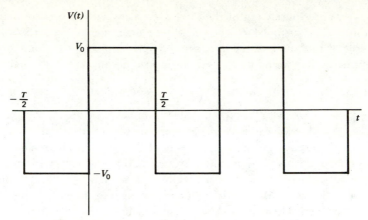

Fig. 5.2 Square wave voltage with period T.

integral into two parts for which $V(t)$ is constant:

$$C_n = \frac{1}{T} \int_{-T/2}^{0} (-V_0)e^{-jn\omega_0 t}\, dt + \frac{1}{T} \int_{0}^{T/2} V_0 e^{-jn\omega_0 t}\, dt$$

$$= \frac{V_0}{jn\omega_0 T}\left(2 - e^{jn\omega_0 T/2} - e^{-jn\omega_0 T/2}\right)$$

Since $\omega_0 T = 2\pi$, the above equation can be written as

$$C_n = \frac{V_0}{2\pi n j}\left(2 - e^{jn\pi} - e^{-jn\pi}\right)$$

With the use of equation 3.24, the above equation becomes

$$C_n = \frac{V_0}{n\pi j}\left(1 - \cos n\pi\right)$$

Note that $\cos n\pi$ is $+1$ for n even $(0, 2, 4, \ldots)$ and -1 for n odd $(1, 3, 5, \ldots)$, so that all the even values of C_n are zero. Any periodic function that when displaced in time by half a period is identical to the negative of the original function:

$$V\left(t \pm \frac{T}{2}\right) = -V(t)$$

is said to have **half-wave symmetry**, and its Fourier series will contain only odd harmonics. The square wave is an example of such a function. If the wave remained at $+V_0$ and $-V_0$ for unequal times, the half-wave symmetry would be lost, and its Fourier series would then contain even as well as odd harmonics.

In addition to its half-wave symmetry, the square wave shown in figure 5.2 is an **odd function**, because it satisfies the relation

$$V(t) = -V(-t)$$

This property arises purely out of the choice of where with respect to the wave the time origin ($t = 0$) is assumed to occur. It is not a fundamental property of the wave. For example, if the square wave in figure 5.2 were displaced by a time of $T/4$, the resulting square wave would be an **even function**, because it would then satisfy the relation

$$V(t) = V(-t)$$

Note that an odd function can have no dc component, since the negative parts exactly cancel the positive parts on opposite sides of the time axis. The cosine is an even function, and the sine is an odd function. Any even function can be written as a sum of cosines ($b_n = 0$ in equation 5.1), and any odd function can be written as a sum of sines ($a_n = 0$ in equation 5.1). Most periodic functions (such as the one in figure 5.1) are neither odd nor even. The Fourier series calculation can often be simplified by adding or subtracting a constant to the value of the function or by displacing the time origin so that the function is even or odd or so that it has half-wave symmetry. One should practice recognizing these three types of symmetries as they occur throughout the remainder of the book.

The odd-numbered coefficients of the Fourier series representation of the square wave are given by

$$C_n = \frac{2V_0}{n\pi j}$$

and the Fourier series is

$$V(t) = \frac{2V_0}{\pi j} \sum_{\substack{n=-\infty \\ n \text{ odd}}}^{\infty} \frac{1}{n} e^{jn\omega_0 t}$$

With the use of equation 3.24 and the fact that $\sin \theta = -\sin(-\theta)$ and $\cos \theta = \cos(-\theta)$, the above equation becomes

$$V(t) = \frac{4V_0}{\pi} \sum_{\substack{n=1 \\ n \text{ odd}}}^{\infty} \frac{\sin n\omega_0 t}{n}$$

The first three terms of the above series ($n = 1, 3, 5$) along with their sum are plotted in figure 5.3. Note that the series, even with as few as three terms, is beginning to resemble the square wave of figure 5.2.

For waveforms more complicated than a square wave, the integrals are more difficult to perform, but it is still usually easier to calculate a Fourier series for a periodic voltage than to solve a differential equation in which the same time-dependent voltage appears. Furthermore, tables of Fourier series for the most frequently encountered waveforms are available and provide a convenient shortcut for analyzing many circuits. Some common waveforms and their Fourier series are listed in figure 5.4.

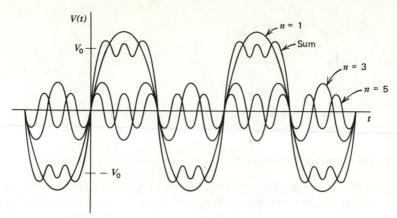

Fig. 5.3 First three terms of the Fourier series for the square wave in Figure 5.2.

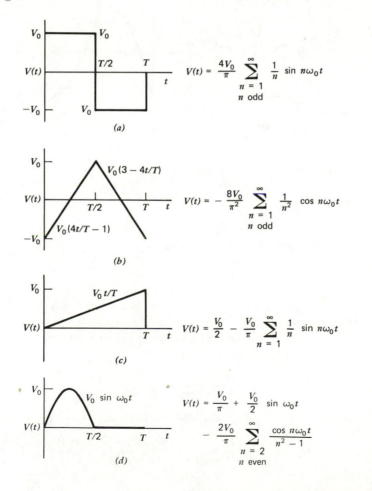

$$V(t) = \frac{4V_0}{\pi} \sum_{\substack{n=1 \\ n \text{ odd}}}^{\infty} \frac{1}{n} \sin n\omega_0 t$$

(a)

$$V(t) = -\frac{8V_0}{\pi^2} \sum_{\substack{n=1 \\ n \text{ odd}}}^{\infty} \frac{1}{n^2} \cos n\omega_0 t$$

(b)

$$V(t) = \frac{V_0}{2} - \frac{V_0}{\pi} \sum_{n=1}^{\infty} \frac{1}{n} \sin n\omega_0 t$$

(c)

$$V(t) = \frac{V_0}{\pi} + \frac{V_0}{2} \sin \omega_0 t$$
$$- \frac{2V_0}{\pi} \sum_{\substack{n=2 \\ n \text{ even}}}^{\infty} \frac{\cos n\omega_0 t}{n^2 - 1}$$

(d)

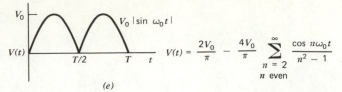

$$V(t) = \frac{2V_0}{\pi} - \frac{4V_0}{\pi} \sum_{\substack{n = 2 \\ n \text{ even}}}^{\infty} \frac{\cos n\omega_0 t}{n^2 - 1}$$

(e)

Fig. 5.4   Fourier series of some common periodic waveforms.

5.2 Square Wave RC Circuit

As an example of how the Fourier series is used to analyze a circuit with a periodic source, consider the series RC circuit in figure 5.5 (a), in which the voltage source is a

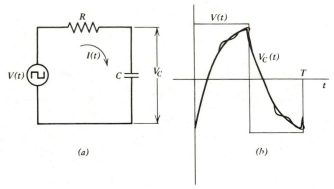

(a)   (b)

Fig. 5.5   A square wave applied to a series RC (a) produces a capacitor voltage as shown in (b). Also shown is the sum of the first three terms of the Fourier series for $V_C(t)$.

square wave. Since the source is periodic, the current $I(t)$ is also periodic with the same period, and it can be written as a Fourier series:

$$I(t) = \sum_{n = -\infty}^{\infty} C'_n e^{jn\omega_0 t}$$

Each C'_n is a phasor current representing one frequency component of the total current in the same way that each C_n represented a component of the phasor voltage in the previous section. The relationship between the two phasors is determined by dividing by the circuit impedance:

$$C'_n = \frac{C_n}{R + 1/j\omega C} = \frac{C_n}{R + 1/jn\omega_0 C}$$

Substituting the value of C_n derived earlier for the square wave gives

$$C'_n = \frac{2V_0}{n\pi(jR + 1/n\omega_0 C)} = \frac{2\omega_0 C(1 - jn\omega_0 RC)V_0}{\pi(n^2\omega_0^2 R^2 C^2 + 1)}$$

for n odd. For n even, C_n' is zero, since C_n is zero for even n. The corresponding current is then

$$I(t) = \frac{2\omega_0 C V_0}{\pi} \sum_{\substack{n=-\infty \\ n \text{ odd}}}^{\infty} \frac{1 - jn\omega_0 RC}{n^2 \omega_0^2 R^2 C^2 + 1} e^{jn\omega_0 t}$$

With the use of equation 3.24, the above current can be written as

$$I(t) = \frac{4\omega_0 C V_0}{\pi} \sum_{\substack{n=1 \\ n \text{ odd}}}^{\infty} \frac{\cos n\omega_0 t + n\omega_0 RC \sin n\omega_0 t}{n^2 \omega_0^2 R^2 C^2 + 1}$$

The voltage across the resistor and capacitor can be determined from the definition of an ideal resistor and an ideal capacitor:

$$V_R(t) = I(t)R = \frac{4\omega_0 RC V_0}{\pi} \sum_{\substack{n=1 \\ n \text{ odd}}}^{\infty} \frac{\cos n\omega_0 t + n\omega_0 RC \sin n\omega_0 t}{n^2 \omega_0^2 R^2 C^2 + 1}$$

$$V_C(t) = \frac{1}{C} \int I(t)\, dt = \frac{4V_0}{\pi} \sum_{\substack{n=1 \\ n \text{ odd}}}^{\infty} \frac{\dfrac{1}{n} \sin n\omega_0 t - \omega_0 RC \cos n\omega_0 t}{n^2 \omega_0^2 R^2 C^2 + 1}$$

The sum of the first three terms of the Fourier series for $V_C(t)$ is shown in figure 5.5(b) for $\omega_0 RC = 1$. For $n\omega_0 RC \gg 1$, this circuit is an integrator, and the voltage across the capacitor is

$$V_C(t) \simeq -\frac{4V_0}{\pi \omega_0 RC} \sum_{\substack{n=1 \\ n \text{ odd}}}^{\infty} \frac{\cos n\omega_0 t}{n^2}$$

which has a shape as shown in figure 5.4(b).

Although the square wave was chosen to illustrate the use of a Fourier series in circuit analysis, circuits with square-wave sources can also be analyzed as transient circuits. During a half period (such as $0 < t < T/2$) when the source voltage is constant, the voltage across the capacitor in figure 5.5(a) has the form

$$V_C(t) = A + B e^{-t/RC}$$

The constants A and B can be determined from

$$V_C(\infty) = A = V_0$$

$$V_C(T/2) = A + B e^{-T/2RC} = -V_C(0) = -A - B$$

The first equation comes from the fact that if the source remains at $+V_0$ forever, the capacitor would charge to voltage V_0. The second equation is required to ensure that the function has half-wave symmetry. The values of the constants are thus

$$A = V_0$$

$$B = -\frac{2V_0}{1 + e^{-T/2RC}}$$

The capacitor voltage is then

$$V_C(t) = V_0 - \frac{2V_0 e^{-t/RC}}{1 + e^{-T/2RC}}$$

for $0 < t < T/2$. The waveform repeats itself for $t > T/2$ with each half cycle alternating in sign. The voltage determined from the above equation is shown in figure 5.5(b) for $\omega_0 RC = 1$.

5.3 Fourier Transforms

Voltages and currents that are not periodic can also be represented as a superposition of sine waves as with the Fourier series, except that instead of a summation over a set of discrete, harmonically related frequencies, the waves have a continuous spectrum of frequencies. A nonperiodic function can be thought of as a periodic function with an infinite period. One must wait forever for the wave to repeat itself. The fundamental angular frequency, which was $\omega_0 = 2\pi/T$ for the Fourier series, approaches zero as the period approaches infinity, and we will represent it as $\Delta\omega$ to remind us that it is an infinitesimal quantity. The various harmonics are separated by the infinitesimal $\Delta\omega$, so that all frequencies are present. If we represent $V(t)$ as a summation, as was done for the Fourier series, we can write

$$V(t) = \sum_{n=-\infty}^{\infty} C_n e^{jn\omega_0 t} = \sum_{n=-\infty}^{\infty} C_n e^{j\omega t} \frac{T\Delta\omega}{2\pi}$$

where we have used the fact that $\omega = n\omega_0$ and $T\Delta\omega = 2\pi$. Since $\Delta\omega$ is infinitesimal, the summation can be replaced with an integral:

$$V(t) = \frac{1}{2\pi} \int_{-\infty}^{\infty} C_n T e^{j\omega t} \, d\omega$$

As before, C_n is given by

$$C_n = \frac{1}{T} \int_{-T/2}^{T/2} V(t) e^{-j\omega t} \, dt$$

However, since T is infinite, we can write

$$C_n T = \int_{-\infty}^{\infty} V(t) e^{-j\omega t} \, dt$$

Although T is infinite, $C_n T$ may be (and usually is) finite. The quantity $C_n T$, which, after integration, is only a function of ω, is called the **Fourier transform** of $V(t)$, and it is written as $\bar{V}(\omega)$. The following two equations are called a **Fourier transform pair**:

$$V(t) = \frac{1}{2\pi} \int_{-\infty}^{\infty} \bar{V}(\omega) e^{j\omega t} \, d\omega \qquad (5.8)$$

$$\bar{V}(\omega) = \int_{-\infty}^{\infty} V(t)e^{-j\omega t}\, dt \tag{5.9}$$

Note the symmetry of the equations. In fact, $\bar{V}(\omega)$ is sometimes defined as $C_n T/\sqrt{2\pi}$ to make the symmetry even more perfect.

Note also that, like the coefficients of the Fourier series (equation 5.6), the Fourier transform $\bar{V}(\omega)$ is generally a complex quantity, unless $V(t)$ happens to be an even function of time. In fact, if $V(t)$ is an odd function of time, the Fourier transform $\bar{V}(\omega)$ is entirely imaginary. Consequently, it is customary when plotting the Fourier transform to plot either its magnitude $|\bar{V}(\omega)|$ or the square of the magnitude $|\bar{V}(\omega)|^2$, called the **power spectrum**, as a function of ω.

As an example of the meaning of the Fourier transform, consider the nonperiodic voltage $V(t)$ given by

$$V(t) = V_0 e^{-t^2/\tau^2}$$

This function is called a **gaussian**, and it is shown in figure 5.6(a). From equation 5.9, the Fourier transform can be calculated by completing the square, with the result:

$$\bar{V}(\omega) = V_0 \int_{-\infty}^{\infty} e^{-t^2/\tau^2 - j\omega t}\, dt$$

$$= V_0 \tau \sqrt{\pi}\, e^{-(\omega\tau/2)^2}$$

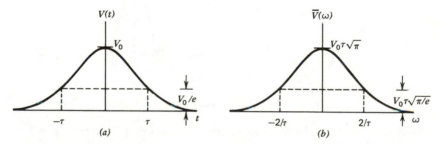

Fig. 5.6   The Fourier transform of a gaussian (a) is another gaussian (b) whose width is inversely proportional to the width of the first.

Note that the Fourier transform of a gaussian happens to be another gaussian, as shown in figure 5.6(b). The gaussian is the only function for which this occurs, but it serves to illustrate an important property of Fourier transforms. The widths of the two curves are related in such a way that when one is narrow, the other is broad, and vice versa. It is a general feature of Fourier transforms that the products of the widths is a number of order unity. The exact value depends on the functions and on how the widths are defined. It is generally true that a circuit that attenuates or amplifies a nonsinusoidal signal without distortion must have a passband at least as wide as the reciprocal of the fastest time variation represented in the signal. To amplify a 1-μsec-wide pulse without distortion requires a circuit with a bandwidth of about 1 MHz. Recall that the RC low-pass filter has a negligible attenuation up to an angular frequency of $1/RC$, and that the circuit responds to an abrupt voltage change in a time of RC, so that the product of the widths is unity.

As another example we will calculate the Fourier transform of the **square pulse** shown in figure 5.7(a) and given by

$$V(t) = \begin{cases} 0 & t < -\tau/2 \quad \text{and} \quad t > \tau/2 \\ V_0 & -\tau/2 \le t \le \tau/2 \end{cases}$$

From equation 5.9 the Fourier transform is

$$\bar{V}(\omega) = V_0 \int_{-\tau/2}^{\tau/2} e^{-j\omega t}\, dt$$

$$= \frac{2V_0}{\omega} \sin \frac{\omega\tau}{2}$$

The magnitude $|\bar{V}(\omega)|$ is plotted as a function of ω in figure 5.7(b). As before, most of the Fourier spectrum is a band of frequencies within about $1/\tau$ of zero.

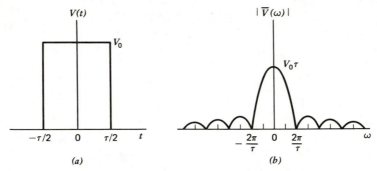

Fig. 5.7 Fourier transform of a square pulse.

It appears that the Fourier transform of even simple functions can be quite complicated. The Fourier transform of a periodic function consists of narrow spikes (called **delta functions**) at harmonically related frequencies. The frequency spectrum of a periodic wave is zero almost everywhere. As with the Fourier series, tables of Fourier transforms are available that greatly simplify the calculations.

Note that the Fourier-transform method is limited to functions that go to zero at large negative and positive times so that the integrals are finite. In practice, this is not a serious limitation, since one can usually integrate to a large but finite time without introducing significant error. A similar technique for analyzing waveforms that start or stop abruptly but continue to infinity in either the positive or negative direction makes use of the **Laplace transform**, in which $e^{j\omega t}$ is replaced with the more general $e^{\alpha t}$ where α is a complex quantity. The Fourier transform then becomes a special case of the Laplace transform in which α is purely imaginary.

5.4 Circuit Analysis with Fourier Transforms

The Fourier transform is used in the analysis of circuits with nonperiodic sources in the same way that the phasor was used for sinusoidal circuits and the coefficients of the Fourier series were used for other periodic circuits. As an example, suppose we wish to calculate the current in a capacitor that has a gaussian voltage as shown in figure 5.6(a) applied across its terminals. Figure 5.8(a) shows the circuit. For this

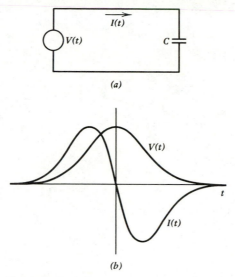

(a)

(b)

Fig. 5.8 For the circuit in (a) a gaussian voltage pulse produces a current as shown in (b).

case, the solution can be written down immediately, without resorting to any Fourier methods:

$$I(t) = C\frac{dV}{dt} = C\frac{d}{dt}\left(V_0 e^{-t^2/\tau^2}\right)$$

$$= -\frac{2CV_0 t}{\tau^2} e^{-t^2/\tau^2}$$

The result is shown in figure 5.8(b). But just for practice, and to illustrate that it really works, we will derive the above result using a Fourier transform. We first write the Fourier transform of $V(t)$, which was calculated in the previous section:

$$\bar{V}(\omega) = V_0 \tau \sqrt{\pi}\, e^{-(\omega\tau/2)^2}$$

From the voltage we can get the Fourier transform of the current using the impedance:

$$\bar{I}(\omega) = \frac{\bar{V}(\omega)}{\bar{Z}(\omega)} \tag{5.10}$$

For the case of a capacitor with a gaussian voltage,

$$\bar{I}(\omega) = j\omega C V_0 \tau \sqrt{\pi} \, e^{-(\omega\tau/2)^2}$$

The current as a function of time is determined using the inverse Fourier transform:

$$I(t) = \frac{1}{2\pi} \int_{-\infty}^{\infty} \bar{I}(\omega) e^{j\omega t} d\omega$$

$$= \frac{jCV_0\tau}{2\sqrt{\pi}} \int_{-\infty}^{\infty} \omega e^{-(\omega\tau/2)^2} e^{j\omega t} \, d\omega$$

If we define a new variable,

$$x = \frac{\omega\tau}{2} - \frac{jt}{\tau}$$

the above integral can be written as

$$I(t) = \frac{2jCV_0}{\sqrt{\pi}\,\tau} e^{-t^2/\tau^2} \left(\int_{-\infty}^{\infty} x e^{-x^2} dx + \frac{jt}{\tau} \int_{-\infty}^{\infty} e^{-x^2} dx \right)$$

The first integral is zero by symmetry, and the second is a frequently encountered integral with a value $\sqrt{\pi}$. The final result is, then,

$$I(t) = -\frac{2CV_0 t}{\tau^2} e^{-t^2/\tau^2}$$

which is the same result derived by simply differentiating the voltage.

The use of Fourier transforms for this particular problem is like cracking a peanut with a sledge hammer. For problems only slightly more complicated, however, the Fourier transform, cumbersome as it is, provides the easiest method of solution. Analyzing a circuit by this method consists of three parts: (1) converting to the frequency domain by calculating the Fourier transform of the sources from equation 5.9; (2) using the circuit impedances to determine the Fourier transform of the unknown from equation 5.10; (3) converting back to the time domain by calculating the inverse Fourier transform of the unknown from equation 5.8. Although the integrals will be difficult, they will usually be less difficult than solving the corresponding differential equation with a time-dependent source.

5.5 Spectrum Analyzers

It is often useful to have a device that will display the Fourier transform $|\bar{V}(\omega)|$ of a voltage as a function of frequency. Such a device is called a **spectrum analyzer**. Suppose we had an ideal filter circuit with a ratio of $|V_{out}/V_{in}|$ given by

$$\left| \frac{V_{out}}{V_{in}} \right| = \begin{cases} 0 & \omega < \omega_0 - \Delta\omega/2 \quad \text{and} \quad \omega > \omega_0 + \Delta\omega/2 \\ 1 & \omega_0 - \Delta\omega/2 \leq \omega \leq \omega_0 + \Delta\omega/2 \end{cases}$$

where ω_0 is a constant that we can adjust and $\Delta\omega \ll \omega_0$. This function is illustrated in figure 5.9(a). From equation 5.8 the output voltage is

$$V_{out}(t) = \frac{1}{2\pi} \int_{-\infty}^{\infty} \bar{V}_{out}(\omega) e^{j\omega t} \, d\omega$$

$$= \frac{1}{2\pi} \int_{\omega_0-\Delta\omega/2}^{\omega_0+\Delta\omega/2} \bar{V}_{in}(\omega) e^{j\omega t} \, d\omega$$

$$\simeq \frac{1}{2\pi} \bar{V}_{in}(\omega_0) e^{j\omega_0 t} \Delta\omega$$

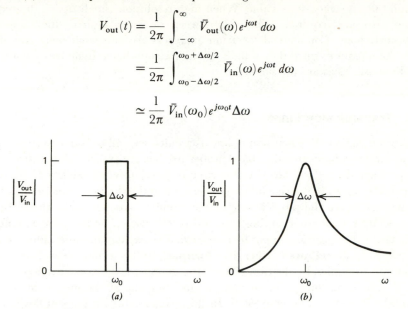

Fig. 5.9 The ideal bandpass filter response in (a) can be approximated by the series RLC circuit whose response is shown in (b).

The magnitude of the output voltage is then proportional to the Fourier transform of $V_{in}(t)$:

$$|V_{out}(t)| = \frac{\Delta\omega}{2\pi} |\bar{V}_{in}(\omega_0)|$$

By measuring $|V_{out}(t)|$ as a function of ω_0, one could then determine the Fourier transform of the input voltage. Such ideal filters are not readily available, however. A reasonable substitute would be a series **RLC** circuit, as shown in figure 4.7. If the output is taken across the resistor (see problem 4.10), the circuit has a bandpass characteristic as shown in figure 5.9(b). In that case, the angular frequency ω_0 is

$$\omega_0 = \frac{1}{\sqrt{LC}}$$

and the bandwidth $\Delta\omega$ is

$$\Delta\omega = \frac{\omega}{Q} = \frac{R}{L}$$

If the angular frequency ω_0 is varied by changing C while keeping R and L constant, the magnitude of the voltage across the resistor is proportional to $|\bar{V}_{in}(\omega_0)|$. If the frequency is automatically swept over the range of interest, the output signal can be displayed on an oscilloscope. One must be careful not to change the frequency too abruptly, however, since the current in the resonance circuit requires a time $\sim 1/\Delta\omega$ to build up its steady-state value. When high resolution (small $\Delta\omega$) is desired, slow sweep rates are required. When high sweep rates are desired, the resolution is necessarily poor. Commercial spectrum analyzers are somewhat more complicated than this situation, since they usually contain a superheterodyne (see section 12.5), but the basic ideas are the same.

5.6 Transmission Lines

Before concluding the discussion of linear circuits, we will consider two examples of linear circuit components that have properties that are rather different from all the components studied so far. In the circuits previously encountered, the circuit elements occurred in discrete lumps connected together by ideal conductors. At high frequencies where the physical size of the circuit is comparable with the distance traveled by a light wave during a period of the wave, the stray capacitance and inductance of the circuit cannot be neglected. The capacitance and inductance here are said to be **distributed** rather than **lumped**. In this section we will consider one important example of a distributed circuit component, the **transmission line**.

Transmission lines come in many forms, but one of the most common is the **coaxial cable** shown in figure 5.10. In the coaxial cable the current flows through

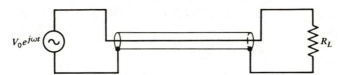

Fig. 5.10 A coaxial transmission line connecting a sinusoidal voltage source to a resistor.

the center conductor and returns in the coaxial outer conductor. One virtue of such an arrangement is that the electric and magnetic fields are confined inside the cable, and so capacitive and inductive coupling to other parts of the circuit are eliminated. The coaxial cable, however, unavoidably has a capacitance per unit length of

$$C' = \frac{C}{l} = \frac{2\pi\varepsilon}{\ln(b/a)} \tag{5.11}$$

where a is the radius of the inner conductor, b is the radius of the outer conductor, and ε is the permittivity of the medium between the conductors. Similarly, the inductance per unit length is

$$L' = \frac{L}{l} = \frac{\mu \ln(b/a)}{2\pi} \tag{5.12}$$

where μ is the permeability of the medium between the conductors.

A lumped circuit representation of the transmission line is shown in figure 5.11. If the line is infinitely long (in the x direction), the impedance as viewed from the terminals on the left can be calculated by removing one of the LC sections, leaving the impedance unchanged, as shown in figure 5.12. This is equivalent to cutting a short

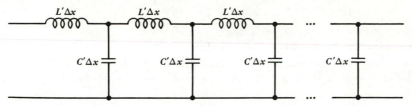

Fig. 5.11 Representation of a transmission line in terms of discrete circuit components.

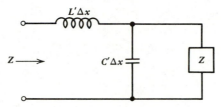

Fig. 5.12 Circuit for calculating the impedance of an infinite transmission line.

piece off the end of the line, which, if the line is infinitely long, still leaves one with an infinitely long line. The input impedance is

$$Z = j\omega L' \Delta x + \frac{Z/j\omega C' \Delta x}{Z + 1/j\omega C' \Delta x}$$

$$= j\omega L' \Delta x + \frac{Z}{1 + j\omega C' Z \Delta x}$$

Solving for Z with Δx small gives

$$Z_0 = \sqrt{L'/C'} \tag{5.13}$$

This is called the **characteristic impedance** of the line. The fact that Z_0 is a real number is quite surprising, because it implies that the line behaves like a resistor despite the fact that it was assumed to have only inductance and capacitance. When a sinusoidal voltage is applied to the line, a current in phase with the voltage flows at the input of the line. The source delivers power indefinitely, but for a line without resistance there is no mechanism for dissipating power. What has happened is that since the line is infinitely long, it can store an unlimited amount of energy. One might object that such an infinite line is unphysical, but note that if the line is finite and terminated with a load resistance equal to Z_0, it will behave as if it were infinite. The ability of a properly terminated transmission line to eliminate the reactance due to stray capacitance and inductance at all frequencies is one quality that makes the transmission line so useful. For a coaxial line, the characteristic impedance can be determined by substituting equations 5.11 and 5.12 into 5.13 to obtain

$$Z_0 = \frac{1}{2\pi} \sqrt{\frac{\mu}{\varepsilon}} \ln\left(\frac{b}{a}\right) \tag{5.14}$$

A typical coaxial line has an impedance of $\sim 50\ \Omega$.

As a sine wave propagates down the line, the phase of the voltage and current will vary with position along the line. If a voltage $V_0 e^{j\omega t}$ is applied at one end of the line, as shown in figure 5.10, the voltage a distance Δx down the line can be calculated using the voltage divider in figure 5.12.

$$V = \left(1 - \frac{j\omega L' \Delta x}{Z_0}\right) V_0 e^{j\omega t}$$

The phase change $\Delta\phi$ is

$$\Delta\phi = \tan^{-1}\left(-\frac{\omega L' \Delta x}{Z_0}\right)$$

Since Δx is small, $\Delta\phi$ can be written as

$$\Delta\phi = -\frac{\omega L' \Delta x}{Z_0} = -\omega\sqrt{L'C'}\ \Delta x$$

The speed with which a wave proceeds down the line is given by

$$v_p = -\frac{\Delta x}{\Delta t} = -\frac{\omega\Delta x}{\Delta\phi} = \frac{1}{\sqrt{L'C'}} \tag{5.15}$$

This is called the **phase velocity**, since it is the rate at which a point of constant phase moves. For the coaxial line the phase velocity is

$$v_p = \frac{1}{\sqrt{\varepsilon\mu}} \tag{5.16}$$

which is just the velocity of light in the medium. If the medium is a vacuum (or air, which has almost the same value of ε and μ), the velocity is

$$c = \frac{1}{\sqrt{\varepsilon_0\mu_0}} = 3 \times 10^8\ \text{m/s}$$

A typical dielectric used in cables is polyethylene, for which the phase velocity is about two-thirds the velocity of light or ~ 20 cm/ns. Stray capacitance and inductance can never be completely eliminated from a circuit. If they could, then a signal would be able to propagate faster than the velocity of light, which is impossible.

Since the phase velocity given by equation 5.15 is independent of frequency, it follows that the various Fourier components of a nonsinusoidal wave will propagate at the same speed down the line, and, on reaching the load, will add together to give a wave of the same shape, except delayed in time. If the phase velocity varies with frequency, the line is said to have **dispersion**, and the shape of the wave would be distorted as it propagated along the line.

If a transmission line is terminated with a resistance other than Z_0 or with an impedance having a reactive (imaginary) component, the impedance at the input will, in general, also have a reactive component and will be a complicated function of the load impedance, the characteristic impedance, and the **electrical length** of the line. Electrical length is a dimensionless number obtained by dividing the line length by the wavelength corresponding to the frequency in use:

$$\frac{l}{\lambda} = \frac{\omega l}{2\pi v_p} \tag{5.17}$$

Note that λ is, in general, different from the free space wavelength, since v_p is usually different from c. Several special cases are worth considering. For a line with a length equal to an integral number of half-wavelengths,

$$\frac{l}{\lambda} = \frac{n}{2} \ (n = 1, 2, 3,...)$$

the magnitude of the voltage and current at the two ends of the line are the same, and the impedance of the load is reflected back to the source without change. For a line with a length equal to an odd number of quarter-wavelengths,

$$\frac{l}{\lambda} = \frac{n}{4} \ (n = 1, 3, 5, ...)$$

the impedance as seen by the source is

$$Z = Z_0^2/Z_L \tag{5.18}$$

The result is reasonable when one considers that a short-circuited transmission line ($Z_L = 0$) must always have zero voltage and maximum current at the shorted end. A quarter-wavelength away the voltage is maximum, and the current is zero. Hence a quarter-wave shorted line looks like an open circuit. Conversely, an open-circuited line ($Z_L = \infty$) will have zero current and maximum voltage at the end. A quarter-wavelength away, the opposite is true, and the line looks like a short circuit. A transmission line can thus be used like a transformer to alter the impedance of a load, but the degree of alteration depends on frequency, unlike an ideal transformer.

For a line of arbitrary length, terminated with an arbitrary impedance Z_L, the input impedance has a more complicated form:

$$Z = Z_0 \frac{Z_L \cos 2\pi l/\lambda + j Z_0 \sin 2\pi l/\lambda}{Z_0 \cos 2\pi l/\lambda + j Z_L \sin 2\pi l/\lambda} \tag{5.19}$$

One should verify that the special cases previously discussed are correctly predicted by equation 5.19.

Whenever a transmission line is terminated with an impedance other than Z_0, a wave propagating down the line will be partially reflected when it reaches the load. The reflected wave adds to the incident wave at every position along the line, producing a **standing-wave pattern** as illustrated in figure 5.13. The ratio of

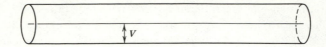

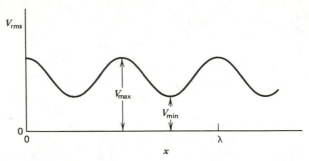

Fig. 5.13 Illustration of standing waves on an improperly terminated transmission line. The *VSWR* is equal to V_{max}/V_{min}.

maximum to minimum rms voltage as a function of position along the line (provided the line is at least a quarter-wave long) is called the **voltage standing wave ratio** (VSWR). The rms current also varies with position, and in fact has the same ratio of maximum to minimum value as the voltage. A *VSWR* of 1 : 1 thus means that the line is properly terminated and there is no reflected wave. A lossless line terminated with either a short or open circuit gives total reflection of the wave and so has an infinite *VSWR* and thus presents a purely reactive load to the source.

Now imagine that a source at one end of a transmission line produces a wave (called the **forward wave**) of voltage V_F. When the wave reaches the opposite end of the line at which a load with $Z_L \neq Z_0$ is connected, a **reflected wave** with voltage V_R appears and propagates back toward the source. From the definition, the *VSWR* is given by

$$VSWR = \frac{V_F + V_R}{V_F - V_R} \tag{5.20}$$

Since the power transported by the two waves is proportional to V^2, the *VSWR* can also be written in terms of the forward and reflected power as

$$VSWR = \frac{\sqrt{P_F} + \sqrt{P_R}}{\sqrt{P_F} - \sqrt{P_R}} \tag{5.21}$$

With a bit of algebra, the ratio of reflected power to forward power becomes

$$\frac{P_R}{P_F} = \left(\frac{VSWR - 1}{VSWR + 1}\right)^2 \tag{5.22}$$

This reflected power is not lost, however, provided the source impedance is properly matched to the input impedance of the transmission line. Rather, it is reflected back again by the source and becomes a part of the forward wave. Note that to achieve an optimum match of the source to the line in such a case requires that the source impedance generally be different from the characteristic impedance of the line. In fact, to achieve maximum power transfer to the load, the source impedance should equal the complex conjugate of the input impedance of the line. This condition is, in general, possible at only a single frequency, unless a very elaborate matching network is employed.

A large VSWR does, however, increase the losses inherent in the line itself. A line without resistance or dielectric losses could tolerate an infinite VSWR without affecting the ability of the source to deliver all its power to the load, provided the breakdown voltage of the line is not exceeded. In a real transmission line, losses occur because of the conductor resistance and dielectric conductivity. The conductor losses tend to increase with the square root of the frequency, and the dielectric losses increase linearly with frequency. Since the mean square voltage and current along the line increase with increasing VSWR for a constant power delivered to the load, the line losses become increasingly serious as the VSWR rises. The attenuation of a transmission line is normally expressed in decibels per unit length for a VSWR of $1:1$, and the variation of attenuation with frequency is usually given. A typical polyethylene coaxial cable with an outside diameter of ~ 5 mm has a breakdown voltage of ~ 2000 V and an attenuation of ~ 1.5 dB/100 ft at 10 MHz, rising to ~ 20 dB/100 ft at 1000 MHz.

The transmission line is a very important component for the circuit designer, especially in circuits that operate at high frequencies. Whenever electrical signals have to be transmitted from one point to another an appreciable fraction of a wavelength away, the inevitable stray capacitance and inductance can lead to quite unexpected and often undesirable results. One is therefore well advised to use a properly terminated transmission line in such a circumstance or at least to use a line of known, constant impedance and length so that the effect of the stray capacitance and inductance can be accurately predicted.

5.7 Waveguides

An interesting extension of the transmission line concept, useful at microwave frequencies ($\sim 10^9 - 10^{12}$ Hz), is the **waveguide**. To understand the operation of a waveguide, consider the parallel plate transmission line shown in figure $5.14(a)$. As in a coaxial line the voltage between the conductors and the current in the conductors varies with position along the line. A voltage that is sinusoidal in time will also vary

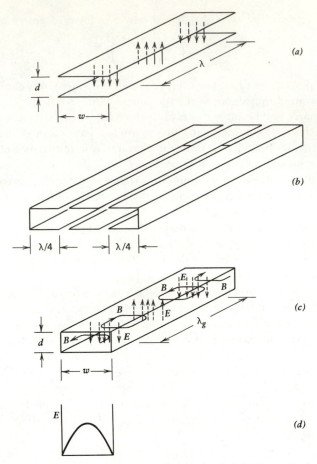

Fig. 5.14 If a parallel plate transmission line (a) is attached to two quarter-wave channels (b), a waveguide configuration results (c) in which the electric field is strongest at the center and falls to zero at the edges (d).

sinusoidally with position along the line with a wavelength λ. If the separation d of the conductors is small compared with their width w, most of the electric and magnetic field energy will reside in the space between the conductors. If the width w is small compared with the wavelength λ, the fields will be nearly constant in a plane perpendicular to the direction of propagation of the wave, but some of the electric and magnetic field will fringe out into the regions surrounding the conductors.

Now, if one wanted to· ensure that the fields are entirely confined between the conductors, use could be made of the fact that a quarter-wave shorted transmission line looks like an open circuit, and two channel-shaped pieces could be attached to the edges of the transmission line, as shown in figure 5.14(b), without greatly affecting the fields near the center of the line. The result, shown in figure 5.14(c), is the basic waveguide configuration. Note that the electric field is essentially the same as in the parallel plate transmission line, except that it goes to zero at the edges, as shown in figure 5.14(d). The shape of the magnetic field is also shown in figure 5.14(c).

The lowest frequency that will propagate in such a rectangular waveguide is called the **cutoff frequency**, and it occurs when the wavelength is twice the width of the guide [corresponding to shrinking the width of the center section of figure 5.14(b) to zero]:

$$f_c = \frac{v_p}{2w} \tag{5.23}$$

For the usual case in which the interior of the guide is empty, the phase velocity v_p is equal to the speed of light, c. For frequencies below cutoff, the waves do not propagate in the guide but decay exponentially with distance along the guide. The wave energy is mostly reflected back to the source. Note that the thickness d does not affect either the shape of the electric fields or the cutoff frequency. The thickness does affect the resistive losses and the power handling capability, however.

Actually, the case considered is only one of an infinite number of **modes** that can propagate in a waveguide. It is called the TE_{10} mode, TE means the *electric* field is everywhere *transverse* (perpendicular) to the direction of propagation. The subscript $_1$ means that the field varies by *one*-half wavelength across the width of the guide. The subscript $_0$ means that there is *no* variation of the electric field in the other direction perpendicular to the propagation of the wave. The TE_{10} mode is important, because it is the mode with the lowest cutoff frequency for a given guide (called the **dominant mode**), and hence it is the mode that allows the smallest guide for a given frequency. TM modes can also be produced in which the *magnetic field* is everywhere *transverse* to the direction of propagation. The waves in an ordinary two-wire transmission line, such as the coaxial cable of the previous section, are TEM waves, since both the electric and the magnetic field are transverse to the direction of propagation. TEM waves can also exist in free space but not in waveguides. Waveguides are usually designed with $d \simeq w/2$ and $f_c \simeq 0.8 f$ so that only the dominant mode will propagate and the attenuation is reduced from the large value that it has near the cutoff frequency. Nonrectangular waveguides are also frequently encountered, and the circular waveguide is a particularly common type.

An alternative description of the waveguide operation is to imagine that the wave, in propagating down the guide, is continually reflected between the side walls of the guide so that the direction of propagation of the wave is always at an angle θ with respect to the axis of the guide, as shown in figure 5.15. For the wave electric field of the dominant mode to go to zero at the boundaries, the angle θ must be such that there is exactly one half-wave across the width w of the guide, so that a crest of the wave and a trough of the wave occur at opposite sides in the same plane along the length of the guide. From figure 5.15 it can be seen that

$$\sin \theta = \frac{\lambda}{2w}$$

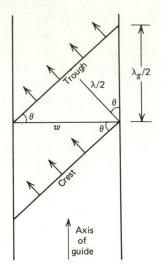

Fig. 5.15 Diagram showing a wave propagating at an angle θ with respect to the axis of a waveguide.

In the direction parallel to the axis of the guide, the distance between crests and troughs of the wave is $\lambda_g/2$, so that

$$\tan \theta = \frac{\lambda_g}{2w}$$

Using the two trigonometric relations,

$$\tan \theta = \frac{\sin \theta}{\cos \theta}$$

and

$$\cos^2\theta = 1 - \sin^2\theta$$

one can solve for λ_g:

$$\lambda_g = \frac{\lambda}{1 - (\lambda/2w)^2} \tag{5.24}$$

The quantity λ_g is called the **guide wavelength**, and it is the wavelength that one would measure along the guide if standing waves were present. Note that λ_g is always longer than λ and that it approaches infinity at the cutoff frequency. For a very large waveguide ($w \gg \lambda$), the guide wavelength is equal to λ, as if the waveguide were absent. Note also that the velocity of a point of constant phase moving parallel to the axis of the guide is greater than the actual phase velocity of the wave by the factor λ_g/λ, so that if the guide is empty ($v_p = c$), the wave appears to propagate down the guide with a velocity greater than the velocity of light. This is a familiar result to anyone who has watched carefully an ocean wave incident on a beach at a slight angle. This result does not violate any principle of physics, because information transmitted down the guide will travel at a slower speed, called the **group velocity**, which must not exceed the velocity of light. Finally, it should be pointed out that since the component of the phase velocity parallel to the axis of the guide is a function of λ, and hence frequency, the waveguide exhibits dispersion, especially near the cutoff frequency, and nonsinusoidal waves propagating down the guide do not maintain their original shape.

One may wonder why a waveguide would be used in place of a coaxial transmission line. The reason is that, at microwave frequencies, the resistance in the conductors of a transmission line and the dielectric, which is required to keep the center conductor out of contact with the outer conductor, produce an unacceptable attenuation over large distances. One may also wonder how a waveguide is connected to an ordinary discrete circuit component such as a resistive load. This is usually done by blanking off the end of the guide and inserting a capacitive stub or inductive loop at the proper place in the guide to couple to the electric or magnetic field.

5.8 Summary

For linear circuits the superposition theorem allows us to calculate the response of a circuit to a nonsinusoidal source by representing the source in terms of sine waves. For a periodic wave the frequencies are discrete and harmonically related. Although there are an infinite number of terms in the Fourier series, the wave can usually be adequately approximated with a few of the lowest frequency components.

For a nonperiodic source a continuous spectrum of frequencies is present. The Fourier transform allows us to calculate this spectrum. Thereafter all calculations are done in the frequency domain, using the impedance of the circuit components. Finally, one must convert back to the time domain, using the inverse Fourier transform. The spectrum analyzer is a device that allows measurement of the Fourier spectrum of a time-dependent voltage.

The transmission line is a linear circuit component that has distributed capacitance and inductance. In addition to being a necessity for carrying high-frequency signals over appreciable distances, transmission lines can also be used as impedance transformers. At microwave frequencies, a form of transmission line called a waveguide can be used, in which the concepts of voltage and current give way almost entirely to the more general description in terms of propagating electric and magnetic fields.

Problems

5.1   Calculate the Fourier series for the periodic voltage shown below:

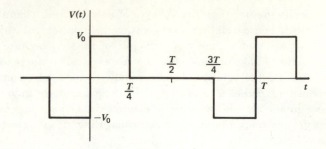

5.2 Calculate the rms value of the voltage in problem 5.1.

5.3 Derive the Fourier series for the waveform shown in figure 5.4(e).

5.4 You have a voltmeter calibrated to read rms voltage, but it responds to the average magnitude of the applied voltage so that it reads correctly only for sine waves. If a square wave voltage is applied to the voltmeter and it reads 1.0 V, what is the rms value of the square wave?

5.5 Show that equation 5.5 is equivalent to equation 5.1 and that equation 5.6 is equivalent to equations 5.3 and 5.4. Calculate C_n in terms of a_n and b_n.

5.6 State which of the waveforms in figure 5.4 are odd, which are even, and which have half-wave symmetry.

5.7 Assume the voltage in figure 5.4(b) is applied at V_{in} in the integrator circuit below in which $R/L = 10^{-3}\omega_0$. Calculate the Fourier series of the output voltage V_{out}.

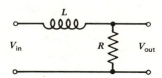

5.8 Suppose a voltage source with voltage as in figure 5.4(d) is connected to an RC low-pass filter. Calculate the value of $\omega_0 RC$ such that the peak-to-peak variation of the lowest (nonzero) Fourier component of the output is 1000 times smaller than the dc component of the output.

5.9 Calculate the Fourier transform $\bar{V}(\omega)$ of the voltage pulse shown below and sketch its magnitude $|\bar{V}(\omega)|$ as a function of ω.

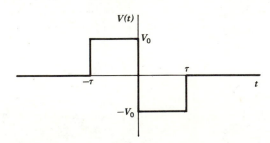

5.10   Use your intuition to sketch as accurately as possible the Fourier transform of the voltage below:

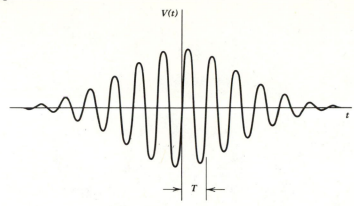

5.11   Write but do not attempt to evaluate the Fourier integral that describes the current in a series RLC circuit with an applied voltage of the form shown in figure 5.7(a).

5.12   Starting with Maxwell's equations (see Appendix D), derive equations 5.11 and 5.12 for the capacitance and inductance per unit length of a circular coaxial cable.

5.13   Calculate the phase shift at 10 MHz in a 3-m length of coaxial cable with $\varepsilon = 2\varepsilon_0$ and $\mu = \mu_0$.

5.14   Calculate the impedance of an infinitely long transmission line having a series resistance R' per unit length.

5.15   Suppose a transmission line of characteristic impedance Z_0 has an electrical length of 3/4 wave and is terminated with a series RL with values R_L and L_L. Show that the impedance at the input looks like a parallel RC circuit, and calculate the values of R and C.

5.16   Calculate the impedance of a 5/8 wave, 100-Ω transmission line that is terminated with a resistance equal to twice the characteristic impedance of the line. Is the impedance inductive or capacitive?

5.17   Starting with Maxwell's equations (see Appendix D), derive expressions for the capacitance per unit length C', the inductance per unit length L', and the characteristic impedance Z_0 of the parallel plate transmission line shown in figure 5.14(a), assuming $d \ll w \ll \lambda$.

5.18   What width and thickness would be desired for propagation of the TE_{10} mode in a rectangular waveguide filled with a dielectric with $\varepsilon/\varepsilon_0 = 9.0$, if the operating frequency is 1000 MHz? Calculate the wavelength λ and the guide wavelength λ_g.

5.19   The group velocity is defined by $v_g = d\omega/dk$ where, for a waveguide, $k = 2\pi/\lambda_g$. Calculate the group velocity for the TE_{10} mode in a waveguide of width w, and show that it is always less than the velocity of light.

chapter 6

Diodes and Rectifiers

6.1 Vacuum Diodes

The remainder of this book will deal primarily with nonlinear devices and circuits. The first and perhaps the simplest nonlinear device that we will consider is the **diode**. We will discuss two types of real diodes and then define an ideal diode in terms of a simple but nonlinear relationship between voltage and current. One should be constantly aware that many of the techniques for analyzing linear circuits are not applicable for circuits with nonlinear components.

The first type of diode we will consider is called a **vacuum diode** and consists of an evacuated tube with two electrodes (hence the name diode), as shown in figure 6.1. The tube is evacuated so that electrons can travel without colliding with

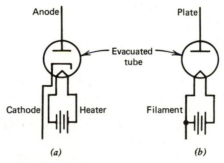

Fig. 6.1 Vacuum diodes. (*a*) Heated cathode. (*b*) Heated filament.

gas molecules. One electrode is called the **cathode**. It is kept at a high temperature by a **heater** or **filament**, which is essentially a resistor that converts electrical energy into heat. The cathode is coated with a material of low **work function** such as barium oxide so that it readily emits electrons. The second electrode is called the **anode** or **plate**. If the anode is positive relative to the cathode, the electrons that boil off the cathode will be drawn to the anode and collected. Hence an electrical current flows from anode to cathode (opposite to the electron flow) in a diode. On the other hand, if the anode is negative relative to the cathode, the electrons are repelled, and no current flows. An alternative version of the vacuum diode makes use of the electrons emitted directly by the filament, as shown in figure 6.1 (b).

The relationship between current and voltage in the vacuum diode is shown in figure 6.2. For small positive voltages the electric field that draws electrons away from

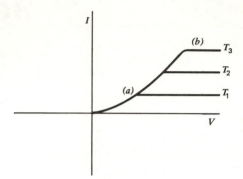

Fig. 6.2 *I* versus *V* characteristic of a vacuum diode for various cathode temperatures. (*a*) Space charge limited region. (*b*) Emission limited region ($T_3 > T_2 > T_1$).

the cathode is partially shielded by the cloud of electrons that surrounds the cathode. This is called the **space charge-limited region**, and the current in amperes is given approximately by **Child's law**:

$$I = 2.33 \times 10^{-6} \, AV^{2/3}/d^2 \qquad (6.1)$$

where A is the area of the cathode in square meters and d is the separation of the cathode and anode in meters. For large positive voltages the electric field is strong enough to collect all the electrons emitted by the cathode, and the current is independent of voltage but depends strongly on the absolute temperature T, as described by **Richardson's equation**:

$$I = 1.2 \times 10^6 \, AT^2 e^{-e\phi/kT} \qquad (6.2)$$

where k is Boltzmann's constant,

$$k = 1.38 \times 10^{-23} \, \text{J/K}$$

ϕ is the work function of the cathode (typically, a few volts, depending on the material), and A is the area of the cathode. This is called the **emission-limited region**. A diode operated in the emission-limited region can be used with a voltage source to produce a good approximation to a current source. The vacuum diode is thus a very nonlinear device.

6.2 *pn* Junction Diodes

Although the vacuum diode was the first widely used diode, for most applications it has been replaced with the **pn junction diode**. The *pn* junction is formed by placing a *p***-type** and an *n***-type semiconductor** (usually silicon or germanium, lightly doped with an appropriate impurity) in contact with one another, as shown in figure 6.3. Actually, the process is not quite so simple as merely placing the semiconductors in contact with one another, since small irregularities in their surfaces would degrade

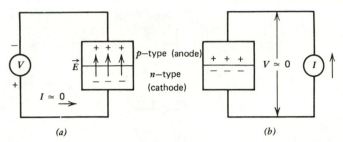

Fig. 6.3 *pn* junction diode. (*a*) Reverse-biased, $I \simeq 0$. (*b*) Forward-biased, $V \simeq 0$.

the quality of the junction. In practice, a *pn* junction is formed by growing a semiconductor crystal with one type of impurity and then abruptly changing to a different impurity while the crystal is still being formed. An *n*-type (negative) semiconductor has a surplus of conduction electrons, and a *p*-type (positive) semiconductor has a deficiency of conduction electrons. The absence of an electron is referred to as a **hole**. In such a junction the *p*-type side is called the anode, and the *n*-type side is called the cathode, by analogy with the vacuum diode.

The appearance of a low-current diode is similar to a resistor — a short cylinder of a few millimeters in diameter, with conducting leads at each end. The cathode is usually marked with a painted band around one end of the cylinder.

If the anode is made negative relative to the cathode, as shown in figure 6.3(*a*), an electric field exists across the junction in the direction shown. This electric field produces a thin layer near the junction which is called the **depletion region**, since it is largely depleted of charge carriers. This depletion region is typically about a micron (10^{-6} meters) thick, but it has a very high resistance and hence opposes the flow of current across the junction. Such a junction is said to be **reverse-biased**, and the current is very small, just as in the vacuum diode with the anode negative relative to the cathode. On the other hand, if a current source is connected to the junction as shown in figure 6.3(*b*), the electrons and holes are pushed towards the junction where they combine, thus causing a current to flow. Such a junction is said to be **forward-biased**, and the voltage across it is quite small.

The current that flows through a *pn* junction as a function of the voltage across the junction is given approximately by

$$I = I_0(e^{eV/kT} - 1) \qquad (6.3)$$

where I_0 is a small constant called the **reverse current**. The quantity kT/e has units of volts and is about 0.026 V at room temperature. The reverse current for a germanium diode is in the microampere (10^{-6} A) range, and the reverse current for a silicon diode is in the picoampere (10^{-12} A) range. The reverse current itself is a sensitive function of temperature. At room temperature, a 10°C increase in temperature will approximately double the reverse current in a germanium diode, and a 6°C increase will approximately double the reverse current in a silicon diode. Figure 6.4 shows the I versus V relation for a germanium and a silicon diode at room

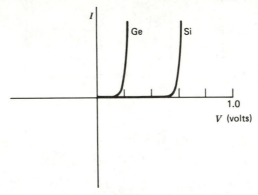

Fig. 6.4 I versus V characteristic for germanium and silicon diode at room temperature.

temperature. Note that the forward voltage drop is relatively constant and is about 0.2 V for a germanium diode and about 0.6 V for a silicon diode. This **forward voltage drop** decreases with increasing temperature. At large values of current, there is an additional voltage drop caused by the resistance of the semiconductor material and its leads. This voltage drop is given by

$$V_{\text{ohmic}} = I r_{\text{ohmic}}$$

so that the total voltage across the terminals of the diode is

$$V = \frac{kT}{e} \ln\left(\frac{I}{I_0}\right) + I r_{\text{ohmic}} \tag{6.4}$$

For most purposes it suffices to neglect the reverse current and to assume the forward voltage drop is constant.

We will now define an ideal diode as a device with the following properties:

$$\left.\begin{array}{ll} I = 0 & \text{If } V < 0 \\ V = 0 & \text{If } I > 0 \end{array}\right\} \tag{6.5}$$

An ideal diode behaves like an open circuit for negative voltages and like a short circuit for positive currents. The symbol for an ideal diode is shown in figure 6.5(a). The arrow points in the direction of current flow.

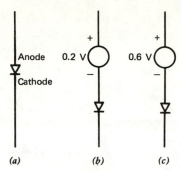

Fig. 6.5 (*a*) Ideal diode. (*b*) Germanium diode. (*c*) Silicon diode.

A germanium and a silicon diode can be approximated as an ideal diode in series with a constant voltage source, as shown in figure 6.5(*b*) and (*c*). It appears from these considerations that the germanium diode is more nearly ideal than the silicon diode. This is not the case, however. Although the silicon diode has a larger forward voltage drop than the germanium diode, it has a smaller reverse current, and as a result the variation of reverse current with temperature is less noticeable. Silicon diodes are normally used in high-current applications, whereas germanium diodes are used for low-voltage applications. An ideal diode cannot dissipate power, since the product *VI* is always zero. A germanium diode dissipates a small power, a silicon diode dissipates about three times as much power at the same forward current, and a vacuum diode dissipates even more power. Although the silicon diode dissipates more power than the germanium diode, it is nevertheless invariably used in high-current applications because it can operate at much higher temperatures (up to $\sim 200°$C) without having an unacceptably high reverse current. By contrast, germanium diodes are worthless above about 85°C.

High-current diodes are often mounted on a **heat sink** to reduce their operating temperature. Real diodes can also be placed in parallel to increase their current-carrying capacity, although care must be taken to ensure that their *V-I* characteristics are closely matched so that the current divides evenly. Alternately, a small resistor can be placed in series with each diode to help equalize the currents. Real diodes also have a maximum allowable reverse voltage called the **peak reverse voltage** (PRV) or **peak inverse voltage** (PIV), above which a large current will flow. When the PRV is exceeded, the diode is usually instantly and permanently destroyed. Diodes typically have a PRV of up to several hundred volts. For higher voltages, diodes can be placed in series, although, again, care must be taken to ensure that the diodes are closely matched so that the reverse voltage divides equally. Alternately, a large resistor can be placed in parallel with each diode to equalize the reverse voltages. With ac voltages, small equalizing capacitors are sometimes used as well to overwhelm any differences in the junction capacitances.

The superiority of the *pn* junction diode over the vacuum diode is readily apparent. It is more nearly ideal; it doesn't require extra power to heat a filament or cathode; it is mechanically rugged, has a much longer life expectancy, and is less costly to manufacture. Nevertheless, vacuum diodes are still sometimes found in equipment such as X-ray machines where very high voltages are involved.

6.3 Rectifier Circuits

One important use of diodes is in circuits that convert ac voltages to dc voltages. Such circuits are called **rectifiers**. Since commercial power lines are usually 60 Hz ac (in the United States) and since most electronic circuits require dc voltages, nearly every electronic device contains diode rectifiers.

The simplest rectifier circuit is the **half-wave rectifier** shown in figure 6.6(a).

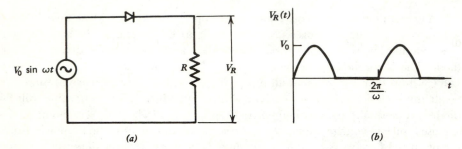

Fig. 6.6 The half-wave rectifier (a) produces an output voltage as in (b).

The diode conducts during the half cycle when the source voltage is positive, and the voltage across the resistive load is the same as the voltage across the source. During the negative half cycle the diode behaves like an open circuit, the current is zero, and the voltage across the resistor is zero. The output voltage for the half-wave rectifier is shown in figure 6.6(b). If the diode were silicon, the peak voltage would be $\sim V_0$ -0.6, and the output voltage would be present for less than half a cycle. The voltage across the load has a dc component given by the average of $V(t)$ over a period:

$$V_{dc} = \frac{V_0}{T} \int_0^{T/2} \sin \omega t \, dt = \frac{V_0}{\pi} \tag{6.6}$$

This result is the same as the zero frequency component of the Fourier series in figure 5.4(d). The half-wave rectifier also has Fourier components at frequencies of ω, 2ω, 4ω, 6ω, and so on. A rectifier is thus nearly always used in conjunction with a low-pass filter to attenuate the ac frequency components (see the next section).

A drawback of the half-wave rectifier is the fact that half of the cycle is missing. This problem is overcome in the **full-wave rectifier** shown in figure 6.7(a), in which a transformer with a center-tapped secondary is used. The diodes alternately conduct for a half-cycle each, producing a current in the load resistor that is always in the same direction. The output voltage for the full-wave rectifier is shown in figure 6.7(b). Note that the peak voltage across the load is $V_0/2$ if a transformer with a

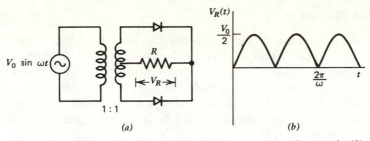

Fig. 6.7 The full-wave rectifier (*a*) produces an output voltage as in (*b*).

$1:1$ turns ratio is used. This is because only half the secondary is used at a time. By using a transformer with a different turns ratio, any output voltage can be achieved. The dc component of the voltage across the load is the same as for the half-wave rectifier, in agreement with the zero-frequency component of the Fourier series in figure 5.4(e). The full-wave rectifier also has Fourier components at frequencies of 2ω, 4ω, 6ω, and so on. The component at frequency ω is missing, and this relaxes the requirements on the low-pass filter that attenuates the ac components.

A circuit that has all the advantages of the full-wave rectifier but which does not require a transformer is the **bridge rectifier** of figure 6.8(a). It does, however, require two extra diodes. During the positive half cycle, the upper-right and the lower-left diodes conduct. During the negative half cycle, the upper-left and the lower-right diodes conduct. The current in the resistor is thus always from right to left, and the voltage across the resistor is as shown in figure 6.8(b). The voltage is

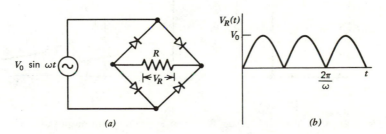

Fig. 6.8 The bridge rectifier (*a*) produces an output voltage as in (*b*).

twice as large as in the full-wave rectifier, but it has the same spectrum of Fourier components. If silicon diodes are used, the peak voltage at the output would be $\sim V_0$ $- 1.2$, since two forward-biased diodes are in series during each half cycle, and there would be brief intervals during which the output voltage remains at zero.

Bridge rectifier units are manufactured with all four diodes internally connected. Such a device is a four-terminal component with one terminal pair for the ac input and a second terminal pair for the dc output.

6.4 Filter Circuits

The main use for rectifier circuits is to convert an ac voltage to a dc voltage. The circuits discussed in the previous section produce voltages with a dc component, but their output also contains ac components. Consequently, a low-pass filter is nearly always used with a rectifier circuit. The lowest (nonzero) Fourier component of the output is usually the largest, and so if we design a low-pass filter that reduces the lowest frequency component to an acceptable level, the higher components will generally be of no concern. It also follows that in a well-designed filter the output will consist of a dc component and a much smaller sinusoidal component with a frequency equal to the lowest Fourier component of the rectified waveform. The sinusoidal part is called the **ripple**, and the percentage ripple is the ratio of the peak-to-peak value of the sinusoidal part to the dc component.

Low-pass filters were discussed in Chapter 4, and it is tempting to use the relations derived there and the Fourier series of Chapter 5 to predict the percentage ripple of various filters. Such a method will often give a reasonably good approximation, but is is usually not precise because of the nonlinear nature of the diodes. The reason is that the output of the rectifier does not look like a voltage source. The resistance is low when the diodes are conducting, but high when they are not conducting. Therefore, the Fourier methods, which were derived for linear circuits, are not strictly applicable.

The simplest filter consists of a capacitor in parallel with the load, as shown in figure 6.9(a). If $\omega RC \gg 1$, the diode conducts only very briefly once each cycle, and

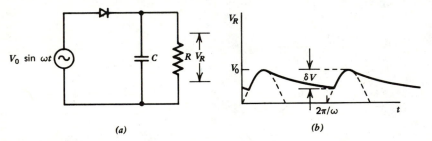

Fig. 6.9 A rectifier with capacitive filter (a) produces an output as in (b).

the voltage thereafter decays exponentially with $\tau = RC$, as shown in figure 6.9(b). For $\omega RC \gg 1$, the dc voltage is $\sim V_0$, and the peak-to-peak ripple voltage is

$$\delta V \simeq V_0(1 - e^{-T/RC}) \simeq \frac{2\pi V_0}{\omega RC} \tag{6.7}$$

With a full-wave or bridge rectifier, the capacitor is recharged twice each cycle, and so the percentage ripple is about one-half as much as with the half-wave rectifier. Equivalently, to achieve the same percentage ripple requires a capacitor only half as large. Notice that the ripple is zero if R is infinite and that it increases linearly with

128 Diodes and Rectifiers

the dc current (V_0/R) in R. It is a general feature of filter circuits that the ripple increases as the dc output current rises. One drawback of this type of filter is that a large **surge** current flows through the source and diode. The same number of coulombs must flow into the capacitor during the brief charging interval as flows out during the much longer discharge interval. The source must then have a very low internal resistance, and the diode must have a large surge current rating (see problem 6.9).

The surge current produced by the capacitive filter can be reduced by adding a resistor in series with the capacitor, as shown in figure 6.10(a). However, such a series

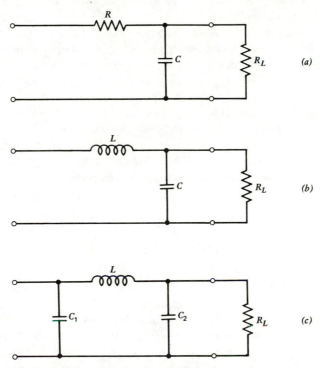

Fig. 6.10 · Some commonly used filters for rectifier circuits.
(a) RC. (b) LC. (c) π-section.

resistor reduces the dc output voltage and wastes power. If the resistor is replaced with an inductor, as in figure 6.10(b), no power is wasted in the filter, but the dc output voltage is still low. By adding a capacitor at the input of the LC filter, as in figure 6.10(c), the dc output voltage is raised, but the surge current reappears. When used with a half-wave rectifier, such a filter provides an attenuation of $1/\omega^2 LC_2$ in addition to that given by equation 6.7, so that the total ripple is

$$\delta V = \frac{2\pi V_0}{\omega^3 R_L C_1 L C_2} \tag{6.8}$$

For a full-wave or bridge rectifier, the dominant Fourier component of the ripple is at an angular frequency of 2ω, so that the percentage ripple is one-eighth of that given by equation 6.8. For this reason, full-wave or bridge rectifiers are usually used when low ripple is desired.

One necessary precaution in the design of LC filters is to ensure that the resonant frequency of the filter is somewhat lower than the lowest Fourier component of the rectifier output ($\omega^2 LC \gtrsim 2$) to avoid resonance effects that might actually enhance the ripple and produce large voltages and currents that could damage the components (see problem 6.10). Since large values of capacitance are desired in these filter circuits, and since the voltages are always of the same polarity, electrolytic capacitors are normally used.

A common feature of all real rectifier/filter circuits is the fact that the dc output voltage varies with the resistance of the load. The voltage **regulation** is expressed as the percentage drop in dc output voltage between the no-load and the full-load conditions. For the same ripple, RC filters have worse regulation than LC filters.

The regulation can be improved by placing a fixed resistor across the output of the power supply in parallel with the load resistor. Although such a resistor will waste power, it will make the output voltage less sensitive to changes in R_L, especially when R_L is large. Such a resistor also serves to discharge the filter capacitors after the ac power is removed, and so it is called a **bleeder resistor**. For applications in which good regulation is required, special **regulator circuits** must be used (see sections 6.7, 8.7, and 9.5).

6.5 Voltage Multiplier Circuits

Sometimes it is useful to have circuits which produce a dc voltage that is higher than the zero-to-peak voltage of the available ac source. Although the usual procedure in such a case is to use a transformer of appropriate turns ratio before the rectifier, an alternate approach is to use a **voltage multiplier circuit**. An example of such a circuit, called a **voltage doubler**, is shown in figure 6.11(a). Such a circuit can be considered as two half-wave rectifiers, each of which charges one of the capacitors to the peak voltage V_0. The capacitors are placed in series so that an output voltage of $2V_0$ is obtained.

A variation of the voltage doubler is the **charge pump circuit** of figure 6.11(b). Diode D_1 charges capacitor C_1 to a dc voltage of V_0 just as with the half-wave rectifier. The voltage across D_1 is thus the sum of the source voltage ($V_0 \sin \omega t$) and the capacitor voltage (V_0). The voltage across D_1 then has a peak value of $2V_0$, and the diode D_2 conducts as required until C_2 is charged to the peak value of $2V_0$.

The circuit is called a charge pump, because during the negative half-cycle the source pumps charge into C_1 through D_1 with D_2 open-circuited, and then during the positive half cycle D_1 becomes an open circuit, D_2 becomes a short circuit, and some of the charge in C_1 flows into C_2. The process continues until enough charge is pumped into C_2 to raise its voltage to $2V_0$. The diodes behave much like the valves in

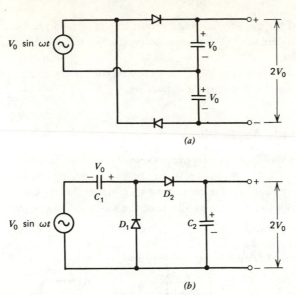

(a)

(b)

Fig. 6.11 Voltage doubler circuits. (a) Conventional doubler. (b) Charge pump.

a water pump, alternately opening and closing each half-cycle. One advantage of the charge pump over the conventional voltage doubler of figure 6.11(a) is that one side of the source and one side of the output are common and hence can be grounded.

The charge-pump concept can be extended to any number of stages, as in the voltage multiplier in figure 6.12. In operation, the capacitors can be regarded as

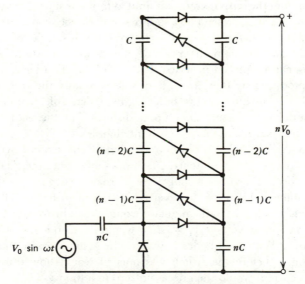

Fig. 6.12 Voltage multiplier circuit.

being in parallel for the ac charging current but in series so far as the dc voltage is concerned. Such circuits can be used for producing dc voltages as high as several hundred kilovolts.

A serious limitation of voltage multiplier circuits is their relatively poor voltage regulation and low output current capability.

6.6 Other Diode Applications

Another use for diodes is in **clipping circuits**, which limit the voltage to some prescribed value. Such circuits are useful for protecting circuit components against damage by overvoltage and for generating special waveforms. Consider the circuit in figure 6.13(a). Whenever the magnitude of the source voltage exceeds V_1, one of the

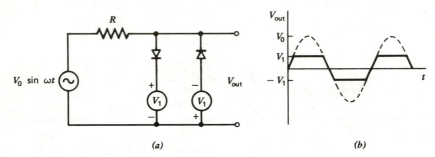

(a)   (b)

Fig. 6.13   The clipping circuit in (a) produces an output voltage as in (b).

diodes conducts, and the output voltage is limited to V_1, as shown in figure 6.13(b). If $V_0 \gg V_1$, the output voltage resembles a square wave. Hence such a circuit is useful for producing square waves. If a voltage $V_1 = 0.2$ or 0.6 V is satisfactory, the dc sources can be omitted, and germanium or silicon diodes used, respectively. For higher voltages, diodes can be placed in series. The output has Fourier components other than the one produced by the source, and such is usually the case with nonlinear circuits. Clipping action often occurs, but is highly undesirable, in audio circuits. The extraneous Fourier components show up as distortion of the audio signal, and in extreme cases they can make the sound unintelligible.

Diodes can also be used to protect circuit components against overvoltage when an inductive load is suddenly switched off. Suppose, for example, that the switch in figure 6.14 has been closed for a long time so that a current $I = V_0/R$ is flowing through the inductor. The diode is reverse-biased and so has no effect. When the switch is opened, the current in the inductor would drop abruptly to zero if the diode were absent, and a large voltage $V_L = LdI/dt$ would develop across the inductor and across the switch. With the diode, however, the same current that was flowing through the source before the switch was opened would flow through the diode afterward. The current would then decay to zero in a time $\tau = L/R$, and the voltage across the inductor and across the switch would never exceed V_0. Such a circuit is

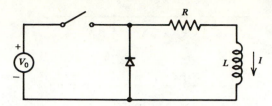

Fig. 6.14 Crowbar circuit in which a diode is used to protect the switch and inductor against a destructive overvoltage when the switch is opened.

called a **crowbar** (cf problem 3.18). The diode would need a PRV rating of V_0 and peak current rating of V_0/R.

Sometimes it is useful to take a periodic ac signal that oscillates between positive and negative values and displace it so that it is either always positive or always negative. Such a circuit is called a **diode clamp**, and it need consist of nothing more than a capacitor and a diode, as shown in figure 6.15(*a*). The circuit is the same as the

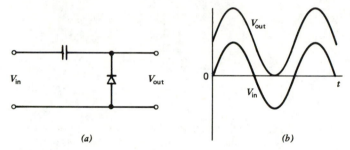

(a)　　　　　(b)

Fig. 6.15 (*a*) Diode clamp circuit which displaces an input voltage so that it is always positive (*b*).

half-wave rectifier except that the capacitor and diode are interchanged. Just as with the half-wave rectifier, the capacitor charges up to a dc voltage equal to the zero-to-peak value of V_{in}. The capacitor is made large enough so that it looks like a short circuit for the ac components of V_{in}. If, for example, V_{in} is a sine wave, V_{out} will equal the sum of V_{in} and the dc voltage on the capacitor as shown in figure 6.15(*b*). Of course, the input voltage need not be a sine wave, but for whatever shape it has, the output voltage will be identical except displaced upward so that it just touches the $V = 0$ axis at its lowest point. By reversing the diode, the input wave can be displaced downward so that it is always negative. Furthermore, by placing a voltage source in series with the diode, the output can be clamped to any desired voltage. Note that the charge pump described in the previous section is a diode clamp plus a half-wave rectifier with a capacitive filter.

A circuit closely related to the diode clamp is the **baseline restoration circuit** shown in figure 6.16(*a*). In the absence of the diode, the circuit is just a low-pass RC

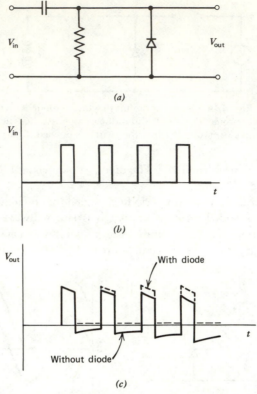

Fig. 6.16 The baseline restoration circuit (*a*) prevents the baseline of an input signal (*b*) from drifting downward (*c*) as the capacitor charges up.

filter such as might be used to observe a small fluctuating voltage superimposed on a larger dc component. Imagine that the signal to be observed consists of a finite string of positive pulses, as shown in figure 6.16(*b*). Although the cutoff frequency of the filter is such as to pass the Fourier components of an individual pulse without causing significant distortion, there is, nevertheless, a low-frequency component arising from the fact that all the pulses are positive and hence do not average to zero. Without the diode, the baseline would thus drift slowly downward in an attempt to eliminate the low-frequency component from the output, as shown in figure 6.16(*c*). In fact, when viewed over a sufficiently long time, the area of the function $V_{out}(t)$ below the axis must exactly equal the area above the axis. Otherwise the capacitor would end up with more charge and hence more voltage than it began with, which would be inconsistent with the dc nature of the circuit at $t = -\infty$ and $t = \infty$. The diode, however, provides a low-resistance path for the capacitor to discharge quickly without causing any current to flow in the negative direction through the resistor. Consequently, each pulse finds the capacitor in the same discharged condition as the previous pulse, and there is no tendency for the baseline to drift downward. If the pulses to be observed were negative, then it would be necessary to reverse the direction of the diode to avoid a corresponding upward shift of the baseline.

As a final example of the many uses of diodes, we will consider how an ac voltmeter or ammeter could be constructed using a D'Arsonval galvanometer. Perhaps the most obvious way to make an ac meter is to use a rectifier and filter to convert the ac to a dc voltage or current which is then connected to a dc meter. Figure 6.17(a) shows perhaps the simplest such circuit. It is just a half-wave rectifier

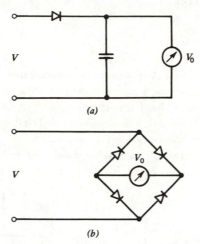

(a)

(b)

Fig. 6.17 Circuits for using a dc voltmeter to measure an ac voltage (a) Peak reading voltmeter. (b) Average reading voltmeter.

with a single filter capacitor. The capacitor will charge up to the peak value of the input voltage. Such a circuit is thus called a **peak-reading ac voltmeter**. If the input is a sine wave, the voltmeter will read a value $V_0 = \sqrt{2} \, V_{rms}$. In addition, the nonideal character of the diode is often important, especially at low voltages. An ac voltmeter of this type would normally have a scale labeled in rms voltage and be calibrated to take into account the factor $\sqrt{2}$ as well as the nonideal character of the diode. One must thus exercise great caution in using such a meter on waveforms that are not sinusoidal (see problem 6.18).

A more usual type of ac meter uses a bridge rectifier and omits the filter capacitor, as shown in figure 6.17(b). If the meter movement has sufficient inertia, it will not respond to the ripple at the bridge output but rather will give a reading proportional to the average magnitude of the applied voltage or current. If the input is a sine wave with a 1-V peak value, the average magnitude is given by

$$\overline{|V|} = \frac{2}{T} \int_0^{T/2} \sin(2\pi t/T) \, dt = \frac{2}{\pi} = 0.637 \text{ V} \tag{6.9}$$

For comparison, remember that the rms value of a 1-V zero-to-peak sine wave is 0.707 V, so that an 11% correction is required in addition to that caused by the nonideal diodes for the meter to read the rms value. Such a meter is called an **average reading ac voltmeter**, and the same precautions apply when a non-sinusoidal voltage is being measured. The same circuit can also be used to read ac current by simply replacing the dc voltmeter with a dc ammeter, while making sure that the diodes have sufficient current-carrying capability. By using operational amplifiers (see Chapter 9), it is possible to construct voltmeters and ammeters which read the **true rms** (TRMS) value.

6.7 Zener Diodes

If the voltage across a reverse-biased *pn* junction is too large, breakdown occurs and the diode conducts. Normally one wants a diode to have a breakdown voltage that is higher than the peak-reverse voltage that the diode encounters in the circuit to which it is connected. On the other hand, a diode with a well-defined, stable, and relatively small, nondestructive breakdown voltage can be a useful circuit element. Such a diode is called a **Zener diode**, **avalanche diode**, or **reference diode**, and its symbol, *V-I* characteristic, and equivalent circuit are shown in figure 6.18. In the

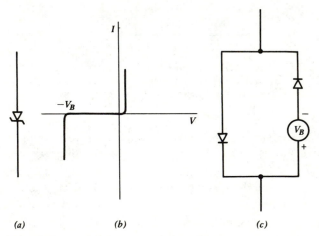

(a)   *(b)*   *(c)*

Fig. 6.18 (*a*) Symbol for ideal Zener diode. (*b*) *V-I* Characteristic of real Zener diode. (*c*) Equivalent circuit for ideal Zener diode.

forward direction a Zener diode behaves like any other diode (i.e., $V \sim 0.6$ V for Si, etc.). In the reverse direction the current increases rapidly when the voltage reaches the breakdown voltage V_B. An ideal Zener diode is like any ideal diode, except that the current goes to minus infinity abruptly at $V = -V_B$. Zener diodes typically have V_B in the range of a few volts to a few hundred volts, although such diodes can be placed in series for higher voltages.

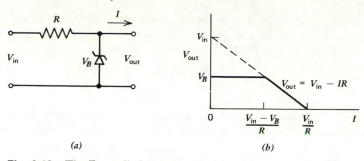

Fig. 6.19 The Zener diode regulator in (a) produces an input voltage as in (b) that is constant so long as the output current is small.

The main use of the Zener diode is as a voltage regulator, as shown in figure 6.19(a). The Zener diode draws as much current as necessary to keep the voltage at V_B, even though the input voltage V_{in} may vary considerably. If too much current flows to the load, $(I > (V_{in} - V_B)/R)$ the voltage will drop below V_B, and the circuit will cease regulating, as shown in figure 6.19(b). For good regulation over a wide range of input voltages and output currents, the input V_{in} must be considerably larger than V_B and/or the resistor R must be small. Unfortunately, both of these remedies results in wasting considerable power in the diode. Unlike an ordinary diode, a Zener diode has a significant simultaneous voltage and current, and so must dissipate power. For the circuit in figure 6.19(a) with $I = 0$, the diode dissipates a power,

$$P = \frac{1}{R}(V_{in} - V_{out})V_{out} \tag{6.10}$$

The Zener diode regulator is closely akin to the clipping circuits previously described.

6.8 Varicap Diodes

Another useful property of real pn junction diodes is the capacitance that appears across the junction when it is reverse-biased. This capacitance arises because of the thin depletion region and is typically in the range of $\sim$10–100 pF. Normally such capacitance would be undesirable in a diode, but sometimes it can be used to advantage. The usefulness of the effect comes from the fact that the capacitance changes as a function of voltage in proportion to $(V + V_0)^{-1/2}$ where V_0 is a positive constant on the order of the forward voltage drop (0.6 V for silicon). The reason for this behavior is that the width of the depletion region increases with increasing reverse voltage. Although any diode will exhibit this behavior, diodes especially made to enhance the effect are called **varicaps** or **varactors**. A typical varicap has a capacitance that varies over about a factor of 10 in the picofarad range. One of the uses for varicaps is in controlling the resonant frequency of an LC circuit by means of a voltage that may be a rapidly varying function of time. If the voltage change δV is small compared with the dc voltage V which is in turn much larger than the constant V_0, the capacitance change is an approximately linear function of δV:

$$\delta C = -\frac{C}{2V}\,\delta V \tag{6.11}$$

Such a voltage-dependent capacitor is a nonlinear circuit element, but it behaves in a nearly linear manner if the ac component of the voltage across its terminals is small compared with the dc component. The nonlinear character of the varicap is often used to advantage in the construction of **frequency multiplier circuits**.

The capacitance of pn junction diodes often becomes significant and poses difficulties when diodes are used at very high frequencies. For such applications **point-contact diodes** are normally used. A point-contact diode has a thin anode wire (called a **catwhisker**) which touches a tiny p-type region formed under the contact on a larger block of n-type semiconductor which serves as the cathode. Such diodes typically have a junction capacitance of $\lesssim 1$ pF, but because of the small area of contact, they are limited to very low currents.

6.9 Summary

A diode is a nonlinear device that conducts current in one direction but not the other. Two common types of diode are the vacuum diode and the pn junction diode. The vacuum diode has a rather complicated V-I characteristic with a space-charge-limited and emission-limited region. The pn junction diode is more nearly ideal. A germanium diode has a forward voltage drop of ~ 0.2 V and a silicon diode has a forward voltage drop of ~ 0.6 V. Both types of diode have a small reverse current.

The main use for diodes is in rectifier circuits that convert ac to dc. Three common rectifier circuits are the half-wave, the full-wave, and the bridge rectifier. To reduce the ripple from a rectifier circuit, some form of filter must be used. Diodes and capacitors can be connected as a voltage multiplier to produce large dc voltages. Diodes can also be used to limit the amplitude of a voltage and to modify waveforms in a variety of ways. A Zener diode is useful as a voltage regulator, and a varicap is useful for controlling the resonant frequency of a circuit by means of a variable voltage.

Because the diode is nonlinear, many of the circuit-analysis techniques previously used must, at best, be applied with considerable caution. There is simply no systematic way to attack a nonlinear circuit in the way that could be done with linear circuits. Each circuit poses a unique problem, and one must usually be content with an approximate solution except in certain cases where Kirchhoff's laws can be applied to the circuit and the resulting nonlinear equations solved exactly.

Problems

6.1   Consider a vacuum diode with a cathode of 1 cm² area, a work function of 2.0 V, and a temperature of 1000 K with an anode 3 mm away. Calculate the emission-limited current, and estimate the voltage at which the transition from emission-limited to space-charge-limited behavior occurs.

6.2   Calculate the reverse current in a germanium and in a silicon diode at room temperature if the forward voltage drop at 100 mA is 0.2 V and 0.6 V, respectively.

6.3   Over what range does the forward voltage drop of a silicon diode vary as the current is varied from 1 mA to 1 A at room temperature, assuming $I_0 = 10$ pA?

6.4   If the diode in figure 6.6(a) is germanium and $V_0 = 1$ V, what is the maximum value of V_R, and over what fraction of the cycle does the diode conduct?

6.5   Calculate the average power dissipated in a 100-Ω load resistor for the half-wave rectifier and for the bridge rectifier, assuming ideal diodes and a sinusoidal source with a peak value of 10 V.

6.6   In the circuit below in which the diode is silicon, sketch the voltage $V_R(t)$, and calculate its maximum value.

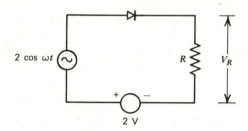

6.7   Describe the symptoms that would result if one of the diodes in the bridge rectifier in figure 6.8(a) failed by becoming an open circuit. What symptoms would result if it failed by becoming a short circuit?

6.8   What minimum PRV rating is required for the diodes in figures 6.6, 6.7, and 6.8 if $V_0 = 100$ V and if a capacitive input filter is used?

6.9   Show that the ratio of the surge current to average current in the diode in figure 6.9 is given by $2\sqrt{\pi\omega RC}$ for $\omega RC \gg 1$.

6.10   If an LC filter is designed with an LC product that is too small, it can actually do more harm than good. For the circuit in figure 6.10(b) with $R_L = \infty$, $L = 1$ H, and $C = 2$ μF, calculate the output ripple voltage if the input consists of a dc component and a 120-Hz, 1-V sinusoidal ac component.

6.11   Suppose you have a power supply that can be represented as a Thevenin equivalent circuit with $V_T = 20$ V and $R_T = 4$ Ω. Calculate the percentage regulation over the range of output currents from 0 to 1 A.

6.12   For the circuit below use reasonable approximations to estimate the dc output voltage, the percentage ripple, and the percentage regulation:

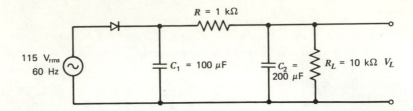

6.13 A bridge rectifier with an RL filter as shown below can be analyzed using Fourier techniques, because two of the diodes are always in conduction. Using the Fourier series of figure 5.4(e), calculate the dc component of the voltage V_R and the value of L required to make the percentage ripple in V_R about 1%.

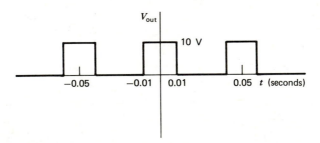

6.14 Suppose in figure 6.11(b) that $C_2 = 10C_1$ and that the source is turned on at $t = 0$ with both capacitors discharged. Sketch the output voltage for the first few cycles of the source.

6.15 Design a clipping circuit using ideal components and a sinusoidal source that will produce a periodic output voltage close to the one shown below:

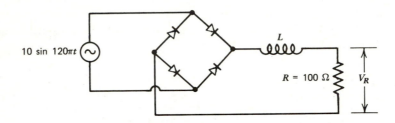

6.16 What is the maximum current that can flow through an ammeter with a 20,000 Ω/V sensitivity and 5000 Ω internal resistance if it is connected in parallel with a silicon diode? Could a germanium diode be used to protect the meter without disturbing its accuracy?

6.17 In the circuit below, the capacitor is initially charged to 100 V and the switch is open. At $t = 0$, the switch is closed. Sketch the voltage across the capacitor and the current through the inductor as a function of time, and show values of voltage, current, and time on your sketch (cf problem 3.18).

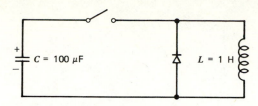

6.18 Suppose you have a peak-reading ac voltmeter, an average-reading ac voltmeter, and a true rms voltmeter, all calibrated for use with sine waves. What rms voltage would each meter indicate if connected to a triangular wave voltage source, as shown in figure 5.4(b) with $V_0 = 100$ V?

6.19 For the circuit below, calculate the power produced by the source and the power dissipated by R, R_L, and the Zener diode.

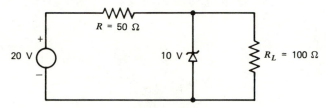

6.20 Design a power supply using real components that will produce a 12-V dc-regulated output over the range 0–100 mA. Use a transformer with a 12.6-V secondary, and calculate what size fuse should be used to protect the 115-V ac primary.

6.21 By what percentage must the voltage V across a varicap be varied in order to vary the frequency of an LC circuit by 1%? Assume the varicap is the only capacitor, and neglect V_0.

chapter 7

Vacuum Tubes and
Field Effect Transistors

7.1 Vacuum Triodes

In this and the next two chapters, we will consider an important class of nonlinear, three-terminal device that is said to be **active**. An active device is one that behaves as if it had internal sources. In this chapter we will consider two such devices, the **vacuum tube** and the **field effect transistor** (FET), which operate on different principles but which behave in a very similar manner.

One example of a vacuum tube has already been considered in the previous chapter where the vacuum diode was described. If one inserts a transparent conducting grid between the cathode and anode (or plate), as indicated schematically in figure 7.1, the device is called a **vacuum triode** (three electrodes). As with

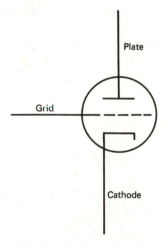

Fig. 7.1 Symbol for a vacuum triode.

the vacuum diode, a heater with its associated voltage source is required to heat the cathode, but since it does not otherwise interact with the rest of the circuit, we will hereafter ignore it. If the grid is made positive relative to the cathode, but not as positive as the plate, it will accelerate the electrons and increase the cathode current, provided the tube is operating in the space-charge-limited region. Some of the electrons are collected by the grid, but if the grid is transparent, many of the electrons will pass through the grid and will be collected by the even more positive plate. A large grid current is usually undesirable, because it requires that the source connected to the grid provide power, and this power must be dissipated by the grid, which in extreme cases may overheat the grid. Consequently, the vacuum triode is normally operated with its grid negative relative to its cathode. In such a case it is energetically impossible for electrons to reach the grid, and so the grid current is always zero. On the other hand, if the grid is sufficiently close to the cathode and not too negative, some of the electrons feel the electric field from the positive plate and pass through the grid and are collected by the plate. In this way a small voltage applied to the grid can be used to control the flow of current between the plate and the cathode. The important property of the vacuum tube is that the source which controls the grid voltage supplies no power, since the grid current is zero, and yet it can significantly alter the power delivered by a source connected between the plate and cathode. The vacuum triode, like other nonlinear, three-terminal, active devices, can thus be used as an amplifier.

Since the vacuum triode is a three-terminal device, its voltage-current characteristic is not as simple as a two-terminal device. If we use the cathode as a reference potential, there are two voltages, grid-to-cathode V_{GC}, and plate-to-cathode, V_{PC}. Since the grid draws no current (assuming $V_{GC} < 0$), the only current is the one that flows from plate to cathode, I_P. With three parameters, one needs a three-dimensional graph to characterize the device completely. However, for convenience, it is customary to plot two of the quantities on a two-dimensional graph, with a family of curves representing various values of the third variable. If I_P is plotted versus V_{PC}, for various V_{GC}, the resulting curves are called the **plate characteristics**. The plate characteristics for a typical vacuum triode in the space-charge-limited region are shown in figure 7.2. The values of the quantities are only representative and may

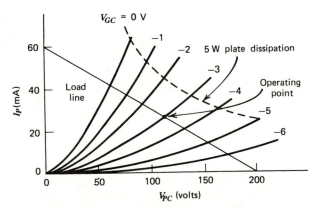

Fig. 7.2 Plate characteristics of a typical vacuum triode showing a representative load line and operating point.

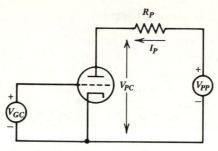

Fig. 7.3 Circuit for determining the operating point of a vacuum triode.

vary by a factor of 10 or more, depending on the construction of the particular tube.

If the triode is connected to a voltage source V_{PP} through a resistor R_P, as indicated in figure 7.3, Kirchhoff's voltage law can be applied to the loop on the right with the result

$$V_{PP} = I_P R_P + V_{PC}$$

Solving for I_P gives

$$I_P = \frac{V_{PP}}{R_P} - \frac{V_{PC}}{R_P} \qquad (7.1)$$

This relation represents a straight line on the plate characteristics, and it is called the **load line**. A typical load line is indicated in figure 7.2 for $V_{PP} = 200$ V, and $R_P = 3.3$ kΩ. Note that the load line intercepts the horizontal axis at V_{PP} and that it has a slope of $-1/R_P$, so that it intercepts the vertical axis at V_{PP}/R_P. In such a circuit the plate current and plate-to-cathode voltage will always lie somewhere on the load line. Their exact value will depend on the grid-to-cathode voltage V_{GC}. The intersection of the load line with the plate current curve corresponding to the appropriate value of V_{GC} is called the **operating point**. The operating point denoted in figure 7.2 assumes a value of $V_{GC} = -3$ V. Notice that the source at the grid in figure 7.3 is labeled with a + toward the grid but that the grid is actually negative relative to the cathode, since $V_{GC} < 0$. Amplifier circuits are normally designed so that the operating point is near the middle of the plate characteristic (i.e., $V_{PC} \sim V_{PP}/2$) so that the largest possible excursions away from the operating point are permitted without causing **saturation** ($V_{PC} \simeq 0$) or **cutoff** ($I_p \simeq 0$).

Care must also be taken to ensure that the product $V_{PC} I_p$ at the operating point is less than the maximum allowed **plate dissipation power** for the particular tube in use. The dashed line in figure 7.2 is a hyperbola showing the limit of allowed operating points for a plate dissipation of 5 W. Operation at points above and to the right of the dashed curve would run the risk of damaging the tube by overheating its plate.

7.2 Triode Linear Equivalent Circuits

Although the vacuum triode is a highly nonlinear device, for small excursions from the operating point the plate characteristics are nearly straight, and the tube behaves in an approximately linear manner. The plate characteristics of figure 7.2 are redrawn in figure 7.4 on a 10-times magnified scale, so that the curvature of the lines is not noticeable.

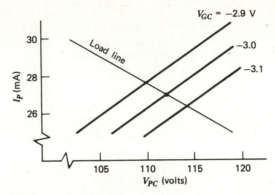

Fig. 7.4 Plate characteristics of figure 7.2 on a magnified scale appear to be linear.

Since the plate current I_P is a function of two variables, V_{GC} and V_{PC}, a small change in either of these quantities will produce a corresponding small change in I_p given by

$$i_P = \left(\frac{\partial I_P}{\partial V_{GC}}\right)_{V_{PC}} v_{GC} \;+\; \left(\frac{\partial I_P}{\partial V_{PC}}\right)_{V_{GC}} v_{PC} \tag{7.2}$$

where a lowercase symbol, i_P, is used to denote an infinitesimal change in a quantity. The first partial derivative is taken holding V_{PC} constant. It is called the **grid-plate transconductance**,

$$g_m = \left(\frac{\partial I_P}{\partial V_{GC}}\right)_{V_{PC}=\text{constant}} \tag{7.3}$$

For a typical triode, the transconductance, which has units of inverse resistance, is in the m℧ range. The second partial derivative in equation 7.2 is taken holding V_{GC} constant (i.e., along the diagonal lines in figure 7.4). Its inverse is called the **plate resistance**,

$$r_P = \frac{1}{\left(\dfrac{\partial I_P}{\partial V_{PC}}\right)_{V_{GC}=\text{constant}}} \tag{7.4}$$

For a typical triode, the plate resistance is in the kilohm range. The value of g_m and r_P

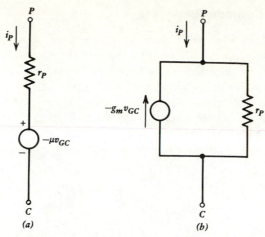

Fig. 7.5 Vacuum tube linear equivalent circuits. (a) Thevenin. (b) Norton.

vary with the location of the operating point as well as with the tube type. Their values are most easily determined by reading them off the plate characteristics of the particular tube after determining the operating point.

Since the vacuum triode behaves in a linear manner for small changes about its operating point, it can be represented by either a Thevenin or a Norton equivalent circuit, as shown in figure 7.5. It will be left as an exercise (problem 7.4) to show that the Thevenin and Norton parameters are given by

$$V_T = -\mu v_{GC} \tag{7.5}$$

$$I_N = -g_m v_{GC} \tag{7.6}$$

$$R_T = R_N = r_P \tag{7.7}$$

where

$$\mu = g_m r_P \tag{7.8}$$

is a dimensionless number called the **amplification factor**. The grid does not appear in the linear equivalent circuits, because no current flows in the grid circuit provided $V_{GC} < 0$.

The vacuum triode behaves as if it had an internal source, and hence it is an active device. This source is different from any encountered so far, however, since its value depends on the value of a voltage elsewhere in the circuit. Such a source is called a **dependent source**. Circuits with dependent sources have unique properties that will be explored in the following chapters. In this case the source voltage depends on the input voltage and provides the necessary coupling between the input and the output, causing amplification.

7.3 Common Cathode Amplifier

The vacuum triode can be used in a circuit such as figure 7.6(a) to amplify a voltage. The analysis of such a circuit always takes place in three parts. First one determines the operating point in the absence of any ac input voltage ($v_{GC} = 0$), as described in the previous section. Then one determines the values of g_m and r_P at the operating point by taking the appropriate derivatives of the plate characteristic curves. Finally, one sets the dc voltages to zero and analyzes the remaining circuit, using one of the linear equivalent circuit representations shown in figure 7.5. The linear equivalent circuit for the amplifier in figure 7.6(a) is shown in figure 7.6(b). It is simply a voltage

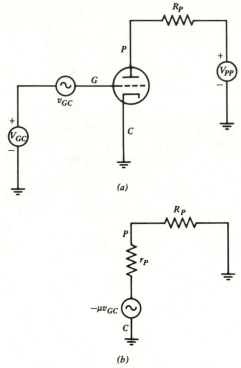

(a)

(b)

Fig. 7.6 The vacuum tube amplifier in (a) can be analyzed using the linear equivalent circuit in (b).

divider, and the ac component of the voltage across R_p is

$$v_{out} = \frac{-\mu v_{GC} R_P}{R_P + r_P}$$

The **amplification** of such a circuit is defined as

$$A = \frac{v_{out}}{v_{in}} \tag{7.9}$$

Like attenuation, amplification is often measured in decibels (see equation 4.15). For the circuit of figure 7.6 with $v_{in} = v_{GC}$, the amplification is

$$A = \frac{-\mu R_P}{R_P + r_P} \tag{7.10}$$

The minus sign indicates that the output is 180° out of phase with the input. Note that if r_P is much less that R_P, the amplification A is just equal to $-\mu$.

The amplifier circuit of figure 7.6 has two drawbacks:

1. Two dc power supplies are required (V_{PP} and V_{GC}), and they must be of opposite sign relative to ground.

2. Both the output and the input have a dc voltage superimposed on them.

These problems are eliminated in the slightly more complicated circuit in figure 7.7(a). In finding the operating point, the capacitors can be treated as open

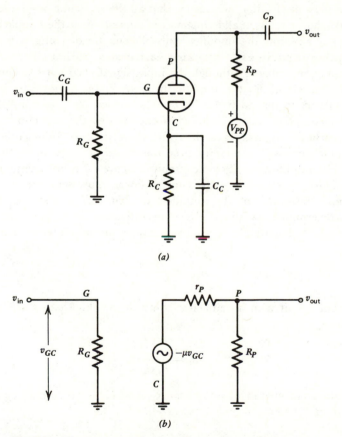

(a)

(b)

Fig. 7.7 The vacuum tube amplifier in (a) can be analyzed using the linear equivalent circuit in (b).

circuits, since only the dc voltages are of significance. The resistor R_G holds the grid at ground potential ($V_G = 0$). Since the grid current is vanishingly small, R_G can be rather large (1 MΩ is typical). This resistor is called a **grid leak**, because it allows the small charge that would otherwise collect on the grid to leak to ground. The resistor R_C allows the cathode to be slightly positive relative to ground by an amount

$$V_C = I_P R_C \tag{7.11}$$

where I_P is the plate (and hence cathode) current at the operating point. Since $V_G = 0$, the dc grid-to-cathode voltage is

$$V_{GC} = -I_P R_C \tag{7.12}$$

Therefore, R_C eliminates the need for the voltage source V_{GC} in figure 7.6(a). Note that finding the operating point requires successive approximations, since V_{GC} depends on I_P, and I_P in turn depends on V_{GC}. Once the operating point is determined, the values of g_m and r_P are determined as usual.

To find the amplification, we assume that all the capacitors are large enough so as to look like short circuits at all frequencies of interest. Since these capacitors couple the ac signals to and from the amplifier while blocking the dc voltages, they are often called **coupling capacitors** or **blocking capacitors**. Such a circuit is limited to amplification of ac voltages, although the input signal need not be sinusoidal. The linear equivalent circuit for the amplifier in figure 7.7(a) is shown in figure 7.7(b). Note its similarity to that in figure 7.6(b). The resistor R_C does not appear in the linear equivalent circuit because it is short-circuited by the capacitor C_C, which is called the **cathode bypass capacitor**. Its value must be sufficiently large that $\omega R_C C_C$ is much greater than unity over the range of frequencies that is to be amplified. The amplification A of the circuit in figure 7.7 is apparently the same as the one in figure 7.6, since it has the same ac linear equivalent circuit.

The input resistance of the amplifier in figure 7.7 is just R_G. The output resistance is determined as with any Thevenin equivalent circuit by setting the sources equal to zero and calculating the resistance between the output terminal and ground:

$$R_{\text{out}} = \frac{R_P r_P}{R_P + r_P} \tag{7.13}$$

The input circuit with its capacitor is just an RC high-pass filter with a 3-dB point given by

$$\omega_C = \frac{1}{R_G C_G} \tag{7.14}$$

If the output is connected to a load resistor R_L, it will also form a high-pass filter with a 3-dB point of

$$\omega_C = \frac{1}{(R_{\text{out}} + R_L) C_P} \tag{7.15}$$

The larger of these two frequencies will determine the lowest frequency for which the amplification is nearly independent of frequency. The capacitor C_P thus eliminates the dc component from the output.

This type of amplifier circuit is called a **common cathode amplifier**, because the cathode is common to both the input and output circuits.

7.4 Cathode Follower Circuit

Consider the amplifier circuit in figure 7.8(a) in which the output is connected to the cathode. Such an amplifier is called a **cathode follower**, since it will turn out that the cathode voltage follows very closely the input voltage. It is also called a **common**

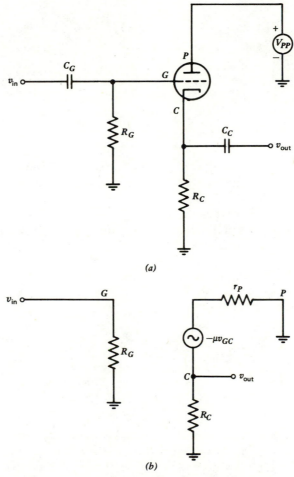

(a)

(b)

Fig. 7.8 The cathode follower circuit in (a) can be analyzed using the linear equivalent circuit in (b).

plate amplifier, because the plate is common to both the input and output circuits. If we assume the capacitors are short circuits for all frequencies of interest, the linear equivalent circuit in figure 7.8(b) is obtained. The ac grid voltage is $v_G = v_{in}$, and the ac cathode voltage is

$$v_C = i_p R_C = \frac{\mu v_{GC} R_C}{R_C + r_P}$$

The grid-to-cathode voltage is then

$$v_{GC} = v_G - v_C = v_{in} - \frac{\mu v_{GC} R_C}{R_C + r_P}$$

Solving for v_{in} gives

$$v_{in} = \left(1 + \frac{\mu R_C}{R_C + r_P}\right) v_{GC}$$

The output voltage is

$$v_{out} = v_C$$

and the amplification is

$$A = \frac{v_{out}}{v_{in}} = \frac{\mu R_C}{r_P + (\mu + 1) R_C} \tag{7.16}$$

For μ very large, A approaches one, and the output closely follows the input. Note that unlike the amplifier previously considered, the output is in phase with the input rather than being shifted 180°.

One might well ask, what use is a circuit that has an amplification of slightly less than one? The answer lies in the input and output resistances. The input resistance is $R_{in} = R_G$. The output resistance is determined by dividing the open-circuit output voltage,

$$v_{out} = A v_{in}$$

by the short-circuit output current,

$$i_{out} = \frac{\mu v_{in}}{r_P}$$

to obtain

$$R_{out} = \frac{A r_P}{\mu} = \frac{A}{g_m} \simeq \frac{1}{g_m} \tag{7.17}$$

Note that this result differs from what would have been obtained had the voltage source been set to zero and the resistance between the output and ground calculated. It is an important property of dependent sources, that they cannot arbitrarily be set to zero, since their value is not constant but depends on a voltage or current elsewhere in the circuit. One can, however, always determine the output resistance of a linear circuit, even with dependent sources, by dividing the open-circuit output voltage by the short-circuit output current. Similarly, the input resistance of any linear circuit, even with dependent sources, can be determined by assuming an input voltage v_{in} and calculating the corresponding input current, i_{in}. The current will be proportional to v_{in}, and the proportionality constant is the input resistance:

$$v_{in} = i_{in}R_{in}$$

Since the input resistance of the cathode follower can be made much larger than the output resistance, the circuit has a large power gain, even through its voltage gain is unity. Such circuits are useful whenever a low resistance load must be driven by a source with a high internal resistance without attenuation. A common example of the use of a cathode follower is to drive a transmission line that is terminated in its characteristic impedance. In such an application, the circuit is sometimes called a **line driver**. The reader should contrast the cathode follower with the transformer, which is also an impedance-matching device but which requires the source to provide all the power that is delivered to the load.

7.5 Grounded Grid Amplifier

The previous two sections dealt with the common cathode and common plate amplifier circuits, respectively. The remaining type of amplifier is the **common grid** or **grounded grid amplifier** shown in figure 7.9(a). As with the previous circuits, the load line is chosen by an appropriate selection of V_{PP} and R_P, and the operating point is determined by the resistor R_C such that $V_{GC} = -I_P R_C$.

The characteristics of the grounded grid amplifier are determined as before by using the ac linear equivalent circuit shown in figure 7.9(b). The input voltage is given by

$$v_{in} = -v_{GC}$$

The output voltage is determined by the voltage divider relation,

$$v_{out} = (-\mu v_{GC} - v_{GC})\frac{R_P}{R_P + r_P}$$

so that the amplification is

$$A = \frac{v_{out}}{v_{in}} = \frac{(\mu + 1)R_P}{R_P + r_P} \tag{7.18}$$

This result is essentially the same (for $\mu \gg 1$) as for the common cathode amplifier, except that the output is in the phase with the input rather than 180° out of phase.

The calculation of the input resistance of the grounded grid amplifier is slightly more difficult. One's first inclination would be to replace the voltage source in figure 7.9(a) with a short circuit and then find the equivalent resistance between v_{in} and ground. However, since the voltage source depends on the value of a voltage

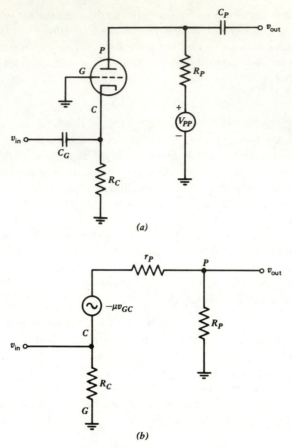

Fig. 7.9 The grounded grid amplifier circuit in (*a*) can be analyzed using the linear equivalent circuit in (*b*).

elsewhere in the circuit, it cannot arbitrarily be set to zero and still permit the measurement of the resistance. Instead, one must return to the basic definition of resistance,

$$R_{in} = \frac{v_{in}}{i_{in}} \tag{7.19}$$

and calculate i_{in} as in figure 7.10, where the circuit of figure 7.9(*b*) has been redrawn, using a Norton equivalent circuit. Applying Kirchhoff's current law to the node at the cathode gives

$$i_{in} = g_m v_{in} + i_1 + i_2$$

$$= g_m v_{in} + \frac{v_{in}}{R_C} + \frac{v_{in} - v_{out}}{r_P}$$

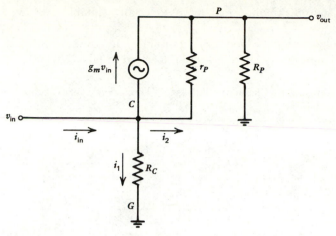

Fig. 7.10 Circuit used for calculating the input resistance of the grounded grid amplifier.

$$i_{\text{in}} = v_{\text{in}} \left(\frac{\mu}{r_P} + \frac{1}{R_C} + \frac{1}{r_P} - \frac{A}{r_P} \right)$$

Substituting the value previously derived for A gives

$$R_{\text{in}} = \left(\frac{1}{R_C} + \frac{\mu + 1}{R_P + r_P} \right)^{-1} \tag{7.20}$$

Unlike the amplifier circuits previously considered, the grounded grid amplifier has a relatively small input resistance $(R_{\text{in}} < R_C)$. The low input resistance makes the grounded grid amplifier especially suitable when the input is driven through a transmission line, which typically has a low characteristic impedance.

The output resistance of the grounded grid amplifier is the same as for the common cathode amplifier:

$$R_{\text{out}} = \frac{R_P r_P}{R_P + r_P} \tag{7.21}$$

In all of the vacuum tube circuits considered, it has been assumed that the input is connected to a voltage source with zero internal resistance and that the output is connected to a load that draws negligible current, such as an ideal voltmeter. If these conditions are not met, the amplification will be correspondingly reduced. One virtue of the vacuum tube amplifier is the almost total isolation of the input from the output, in the sense that the input resistance is not affected by the load connected to the output and the output resistance is not affected by the resistance of the source connected to the input. The grounded grid amplifier is an exception to this rule, however, and the isolation is not perfect (see problem 7.12). However, the grounded grid amplifier does reduce to a bare minimum the capacitive coupling between the input and output circuits, making it especially suitable for use at very high frequencies.

The voltage amplification, input resistance, and output resistance of the three types of vacuum tube amplifiers are summarized in table 7.1.

TABLE 7.1   Characteristics of the Three Types of Vacuum Tube Amplifiers

	Common Cathode	Cathode Follower	Grounded Grid
Amplification (voltage) $A =$	Large $-\dfrac{\mu R_P}{R_P + r_P}$	Small $\dfrac{\mu R_C}{r_P + (\mu + 1)R_C}$	Large $\dfrac{(\mu + 1)R_P}{R_P + r_P}$
Input resistance $R_{\text{in}} =$	Large R_G	Large R_G	Small $\left(\dfrac{1}{R_C} + \dfrac{\mu + 1}{R_P + r_P}\right)^{-1}$
Output resistance $R_{\text{out}} =$	Medium $\dfrac{R_P r_P}{R_P + r_P}$	Small $\dfrac{R_C r_P}{(\mu + 1)R_C + r_P}$	Medium $\dfrac{R_P r_P}{R_P + r_P}$

7.6   Multigrid Tubes

For a vacuum triode the amplification factor μ seldom exceeds about 100. A larger amplification and a lower capacitive coupling between the grid and plate can be achieved by adding a second grid, called the **screen grid**, between the control grid and the plate, as shown in figure 7.11(a). Such a tube is called a **tetrode** (four

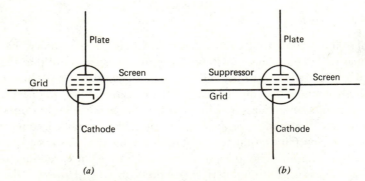

(a)   (b)

Fig. 7.11   Multigrid tubes. (a) Tetrode. (b) Pentode.

electrodes). The screen grid is usually held at a constant voltage, intermediate between the cathode and plate voltage. Its function is to reduce the dependence of plate current on plate-to-cathode voltage by providing a nearly constant electric field to accelerate electrons away from the cathode. Most of the electrons that pass through the screen are collected by the plate, but some are collected by the screen and contribute to a small screen-grid current.

One drawback of the tetrode is that whenever the plate becomes negative relative to the screen, secondary electrons that are knocked off the plate by the incident primary electrons are attracted to the screen. This causes an undesirably high screen current and can cause the plate current to reverse direction. To eliminate this effect, a third grid, called the **suppressor grid**, is inserted between the screen and the plate, as shown in figure 7.11(b). Such a tube is called a **pentode** (five electrodes). The suppressor grid is usually held at a constant voltage near that of the cathode and, in fact, is often connected internally to the cathode. The suppressor-grid current is very small, and it effectively repels secondary electrons and drives them back to the plate.

Typical plate characteristic curves for a pentode with a constant screen voltage are shown in figure 7.12. Note that the tube behaves very much like a current source,

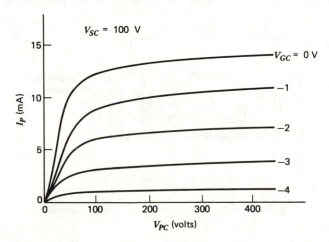

Fig. 7.12 Plate characteristics of a typical pentode vacuum tube.

except at low voltages. The plate resistance r_P is quite high (megohms), and the amplification factor μ is also high ($\sim 10^4$). Either linear equivalent circuit in figure 7.5 can be used to represent the pentode, but the Norton equivalent circuit is more realistic, and r_P is often sufficiently large that it can be omitted entirely without introducing significant error.

7.7 Junction Field Effect Transistors

A semiconductor device with characteristics very similar to the pentode vacuum tube is the **field effect transistor** (FET) shown in figure 7.13. The device consists of a narrow channel of n-type silicon semiconductor sandwiched between two pieces of p-type silicon. The two p-type sides are connected together and are called the **gate**. One end of the channel is called the **source**, and the other end is called the **drain**. If the gate is negative relative to the source and drain, it forms a reverse-biased pn junction, and very little current flows in the gate as with the grid of a vacuum tube. The reverse bias at the junction forms a depletion region that extends into the channel and is thickest near the drain. The width of the resulting channel and hence the current flow from the drain to the source is thus controlled by the voltage applied to the gate. The gate controls the flow of electrons from source to drain in much the same way as the grid controls the flow of electrons from cathode to anode in the vacuum tube. Unlike the vacuum tube, the FET can be made in either polarity, since there are two types of charge carriers, electrons, and holes. A p-channel sandwiched between two n-type semiconductors behaves in an analogous fashion, provided the signs of all the voltages and currents are reversed. The schematic symbols for these devices are shown in figure 7.14. The arrow at the gate is drawn in the direction of current flow by analogy with the symbol for a diode, but since the gate is normally reverse-biased, what little current does flow in the gate actually flows opposite to the direction of the arrow.

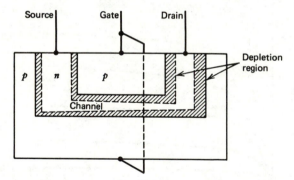

Fig. 7.13 n-channel junction field effect transistor.

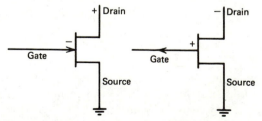

Fig. 7.14   Symbols for junction field effect transistors. (a) n-channel. (b) p-channel.

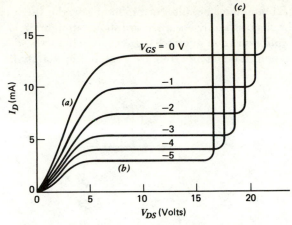

Fig. 7.15 Drain characteristics for a typical n-channel JFET. (a) Ohmic region. (b) Pinch-off region. (c) Breakdown region.

The drain characteristics of a typical n-channel FET are shown in figure 7.15. At low values of V_{DS}, the drain current varies nearly linearly with V_{DS}. This is appropriately called the **ohmic region**. In the **pinch-off** region, I_D depends strongly on V_{GS} but only weakly on V_{DS}. Eventually, breakdown occurs, and a large current flows in a manner reminiscent of the Zener diode. The drain characteristics of the FET are similar to the plate characteristics of a pentode vacuum tube, except that the voltages are usually somewhat smaller.

An FET circuit is analyzed in the same way as a vacuum tube circuit. The operating point is first determined by considering the dc circuit. Normally the operating point is chosen near the middle of the pinch-off region. At the operating point the value of the **forward transconductance**,

$$g_{fs} = \left(\frac{\partial I_D}{\partial V_{GS}}\right)_{V_{DS} = \text{constant}} \tag{7.22}$$

and the output resistance,

$$r_{os} = \cfrac{1}{\left(\cfrac{\partial I_D}{\partial V_{DS}}\right)_{V_{GS} = \text{constant}}} \tag{7.23}$$

can be determined. The subscript s indicates that the device is being used in the common source configuration. Then the linear equivalent circuit can be used in which all the dc sources are turned off and the Norton equivalent current is $-g_{fs}v_{GS}$ and the Norton equivalent resistance is r_{os}, as shown in figure 7.16. Often the output resistance r_{os} is so large that it can be taken as infinite, in which case the linear equivalent circuit for the FET becomes nothing more than a current source $-g_{fs}v_{GS}$ between the source and the drain.

The basic FET amplifier circuits corresponding to the three types of vacuum tube amplifier circuits previously discussed are shown in figure 7.17. The **common source**, **source follower**, and **grounded gate** have characteristics identical to the corresponding vacuum tube circuits and are analyzed in exactly the same manner.

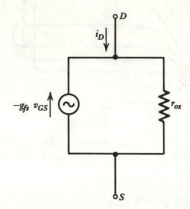

Fig. 7.16 FET linear equivalent circuit.

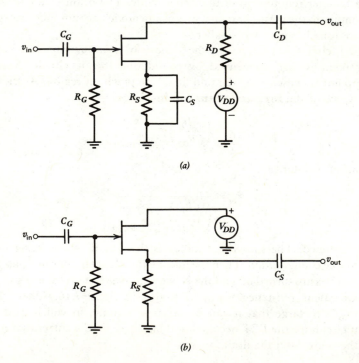

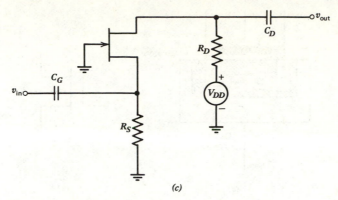

Fig. 7.17 FET amplifier circuits. (*a*) Common source. (*b*) Source follower. (*c*) Grounded gate.

7.8 Insulated-Gate Field Effect Transistors

The device described in the previous section is called a **junction field effect transistor** (JFET) because the gate forms a reverse-biased junction with the channel. The input resistance of the gate is typically $\sim 10^9\,\Omega$. An even higher input resistance can be achieved by placing a thin insulating layer between the gate and the channel. Such a device is called an **insulated-gate field effect transistor** (IGFET). The most common type of IGFET uses a metal oxide such as SiO_2 as an insulator and is called a **metal oxide semiconductor field effect transistor** (MOSFET). In this way, the input resistance of an FET can be increased to $\gtrsim 10^{14}\,\Omega$.

The gate-to-channel capacitance is also very small (a few picofarads) so that a very small electrical charge applied to the gate can result in voltages large enough to destroy the FET. Great care must be taken when handling and installing a MOSFET in a circuit to avoid the buildup of static electricity on its gate. Some MOSFETs have a pair of built-in, back-to-back Zener diodes between the gate and source to prevent damage by overvoltage.

MOSFETs are made in two types, called the **depletion type** and the **enhancement type**, as shown in figure 7.18. In the depletion type the channel is open when the gate-to-source voltage is zero. In the enhancement type the channel is normally closed but can be opened by forward-biasing the gate. The depletion type has the advantage that gate biasing is especially simple, since a dc gate-to-source voltage of zero often provides a quite acceptable operating point. On the other hand, the enhancement type requires a gate voltage of the same polarity as the drain voltage, and so biasing can be obtained by a voltage divider from the same power supply. This provides the possibility of dc coupling two or more FET amplifier stages. An enhancement-type device is also required whenever a zero drain current is required at zero gate voltage.

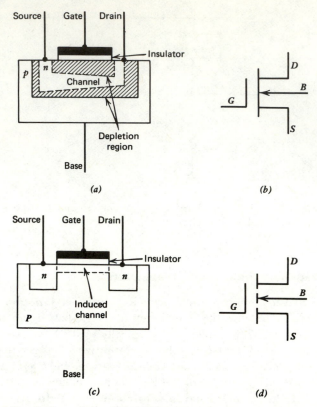

(a)

(b)

(c)

(d)

Fig. 7.18 *n*-channel MOSFET's. (*a*) Depletion-type. (*b*) Symbol. (*c*) Enhancement-type. (*d*) Symbol.

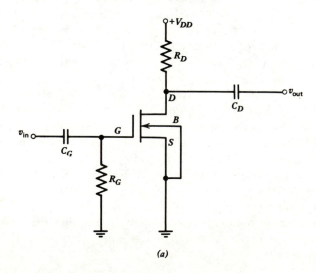

(a)

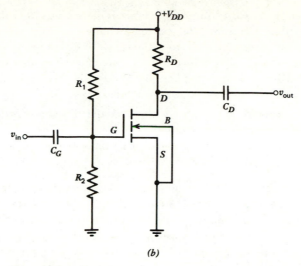

(b)

Fig. 7.19 MOSFET common collector amplifiers. (a) Depletion type with zero bias. (b) Enhancement-type with positive bias.

The gate electrode is insulated from the channel, and the remainder of the gate is called the **base** and is brought out on a separate lead. The MOSFET is thus actually a four-terminal device. The base is often used to determine the operating point while the incremental ac signal is applied to the gate. Just as with the JFET, the MOSFET can also be made with a p-channel, in which case all the voltages and currents are reversed from the n-channel.

Figure 7.19 shows two examples of common source MOSFET amplifiers. Figure 7.19(a) uses a depletion-type MOSFET with zero gate bias. Figure 7.19(b) uses an enhancement-type MOSFET with positive gate bias obtained from a voltage divider connected to the positive dc voltage supply V_{DD}.

Compared with the vacuum tube, the FET is mechanically rugged but electrically fragile (easily destroyed by overvoltage or current), is physically smaller, operates at lower voltages, requires no heater power, and lasts forever if not abused.

7.9 Summary

The vacuum tube and the FET are two important, three-terminal, nonlinear, active devices having similar characteristics that enable them to be used as amplifiers. The analysis of such circuits takes part in three stages:

1. The operating point is determined from the dc circuit by drawing the load line and finding its intersection with the appropriate grid (or gate) voltage curve on the plate (drain) characteristics. All capacitors are treated as open circuits.

2. The grid-plate (forward) transconductance is determined from the incremental change in plate (drain) current produced by a change in grid-to-cathode (gate-to-source) voltage with constant plate-to-cathode (drain-to-source) voltage, and the plate (output) resistance is determined from the incremental change in plate (drain) current produced by a change in plate-to-cathode (drain-to-source) voltage with constant grid-to-cathode (gate-to-source) voltage.

3. The dc sources are then turned off, and the Thevenin or Norton linear equivalent circuit representation with parameters calculated above are used to calculate the circuit behavior in the presence of a small ac input signal. All capacitors are treated as short circuits, provided their values are sufficiently large.

Although the tube and FET are nonlinear devices, for small signals they behave in a nearly linear manner. For many cases the plate (output) resistance is sufficiently large that the equivalent circuit of the device consists of nothing more than a single current source with a value proportional to the ac input voltage.

The FET and the bipolar transistor to be discussed in the next chapter have replaced the vacuum tube in all but a few highly specialized applications such as circuits that employ a combination of high voltage, high power, and high frequency.

The development of active semiconductor devices in the 1950s caused a revolution in the field of electronics and gave birth to a new generation of electronic gadgets of ever-increasing sophistication. Even today, the rate of development of new semiconductor devices and circuits is so staggering that one is left to contemplate how society will be altered by the epoch of electronics which has dramatically overtaken us.

Problems

7.1 For the plate characteristics in figure 7.2, estimate the value of I_P and V_{PC} at the operating point, assuming $V_{PP} = 150$ V, $R_P = 2.5$ kΩ, and $V_{GC} = -2$ V.

7.2 Estimate the value of g_m, r_P, and μ at the operating point in problem 7.1.

7.3 Consider the vacuum tube circuit in figure 7.3 with an operating point as shown in figure 7.2. (a) If V_{GC} is made more negative, does r_P increase or decrease? (b) If R_P is increased, does g_m increase or decrease? (c) If V_{PP} is increased, does r_P increase or decrease?

7.4 Show that the Thevenin and Norton parameters in figure 7.5 are consistent with equations 7.5 to 7.7.

7.5 In the circuit in figure 7.7 (a), the vacuum tube has $\mu = 5000$ and $r_P = 10^6$ Ω at its operating point of $I_P = 20$ mA and $V_{GC} = -1.0$ V. If the amplifier is to have an input resistance of 10^5 Ω and an output resistance of 10^4 Ω, what should be the values of R_P, R_C, and R_G? What is the amplification A of the circuit (assuming the capacitors are short circuits for ac)? What value of C_G would give a low-frequency 3-dB point of 100 Hz?

7.6 Find the operating point for the circuit in figure 7.7(a), using the plate characteristics in figure 7.2, assuming $R_P = 1700\ \Omega$, $R_C = 100\ \Omega$, and $V_{PP} = 100\ \text{V}$.

7.7 Calculate the amplification A for the circuit in figure 7.7(a) if $C_C = 0$. The omission of the cathode capacitor provides **negative feedback** which has numerous desirable properties to be discussed in subsequent chapters. Show that for μ sufficiently large, A is independent of any of the properties of the tube.

7.8 Calculate the amplification A for the circuit in figure 7.7(a) if the output is connected to a load resistor R_L.

7.9 Suppose that a small capacitance C exists between the grid and the plate of a vacuum tube amplifier, as shown below. Show that the input capacitance of the amplifier is given by $C_{in} = (1 + |A|)C$, where A is the amplification. This enhancement of the input capacitance is called the **Miller effect**. When a low-input capacitance is required, a tetrode or pentode is normally used, because the screen grid greatly reduces the value of C.

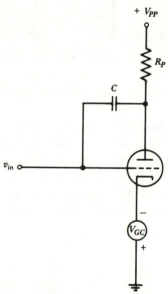

7.10 Calculate the value required for R_C in the cathode follower circuit of figure 7.8(a) if $V_{PP} = 100\ \text{V}$ and $V_{GC} = -2\ \text{V}$, using the plate characteristics in figure 7.2.

7.11 Estimate the amplification A for the circuit of figure 7.8(a) for the conditions given in problem 7.10.

7.12 Calculate the input and output resistance of the grounded grid amplifier in figure 7.9(a), assuming $\mu = 99$, $r_p = 10\ \text{k}\Omega$, $R_P = 10\ \text{k}\Omega$, and $R_C = 100\ \Omega$. By what percentage does the input resistance change if a 10-kΩ load resistor is connected to the output? By what percentage does the output resistance change if the resistance of the source connected to v_{in} is 100 Ω?

7.13 Calculate numerical values for the amplification, input resistance, and output resistance, for each of the three types of amplifiers listed in table 7.1, assuming $R_P = 10$ kΩ, $R_C = 100$ Ω, $R_G = 1$ MΩ, and a pentode vacuum tube with $g_m = 0.01$ $\mho$ and $r_p = 1$ MΩ.

7.14 Calculate the amplification A for the circuit below using the plate characteristics in figure 7.12.

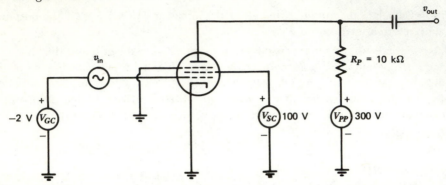

7.15 Sketch what v_{out} would look like as a function of time for the circuit of problem 7.14 if $v_{in} = 10 \sin \omega t$.

7.16 Calculate the amplification A for the circuit in figure 7.17(a), assuming $g_{fs} = 2.5$ m$\mho$ and $r_{os} = \infty$.

7.17 Calculate the values required for R_G, R_D, and R_S in the circuit in figure 7.17(a) if the FET has $g_{fs} = 0.01$ $\mho$ and $r_{os} = \infty$ and the amplifier is to have an input resistance of 10^6 Ω, an output resistance of 10^3 Ω and an operating point with $I_D = 10$ mA and $V_{GS} = -1$ V. The capacitors can be considered short circuits for ac.

7.18 In the circuit below, the FET has $g_{fs} = 10^{-3}$ $\mho$ and $r_{os} = \infty$ at its operating point. Calculate the amplification A. You may assume the capacitors act as short circuits for ac.

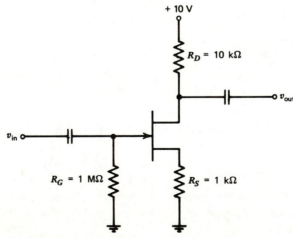

7.19 For the circuit below in which $g_{fs} = 0.04\ \mho$ and $r_{os} = \infty$, calculate $A_1 = v_1/v_{in}$ and $A_2 = v_2/v_{in}$, assuming the capacitors are short circuits for ac.

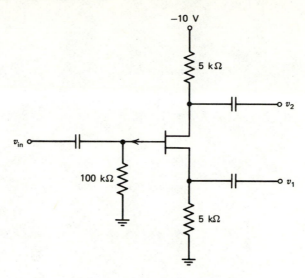

7.20 In the circuit below, calculate the dc gate-to-source voltage V_{GS} if the FET is biased such that $V_{DS} = 0.5\ V_{DD}$.

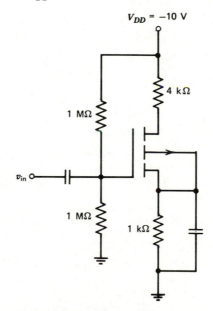

7.21 Calculate the amplification A, the input resistance R_{in} and the output resistance R_{out} for each of the circuits in figure 7.17 in terms of g_{fs}. Assume $r_{os} = \infty$ and the capacitors are short circuits for ac.

chapter **8**

Bipolar
Transistors

8.1 Construction and Operation

The FET described in the previous chapter is but one example of an active semiconductor device. The first such device to be invented was the **bipolar transistor**, and it remains the most common of the active semiconductor devices. In many ways it resembles the vacuum tube and FET, but it also has important differences, and in some ways it is simpler than those devices. The bipolar transistor is so named because current is carried simultaneously by charges of both polarities (electrons and holes) rather than by a single species, as in the FET which is an example of a **unipolar** device.

The bipolar transistor can be made by placing a thin p-type semiconductor in a sandwich between two n-type semiconductors, as shown in figure 8.1. Actually, a

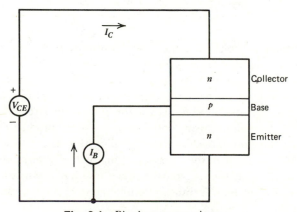

Fig. 8.1 Bipolar npn transistor.

transistor is manufactured by abruptly changing the doping material twice while the semiconductor crystal is being grown. If a current I_B flows into the base, the base-to-emitter junction is forward-biased, and it behaves like any forward-biased pn junction; that is, the voltage across the junction is small (~ 0.2 V for germanium and ~ 0.6 V for silicon). If a large, positive voltage V_{CE} is applied between the collector and emitter, the collector-to-base junction is reverse-biased, and the collector current I_C is small. However, since the base is very thin, most of the electrons that flow from the emitter to the base diffuse across the narrow base region before they have a chance to recombine with a hole and are collected by the positive collector in much the same way that electrons that pass through the grid of a vacuum tube are attracted to the plate. In this way, a small base current is capable of controlling a much larger collector current. The bipolar transistor thus resembles the vacuum tube and FET, except that the input (base-to-emitter junction) is biased so as to look like a short circuit rather than an open circuit.

Like the FET, the bipolar transistor comes in two types which are identical except that the sign of the voltages and currents are reversed. These are called the *npn* and the *pnp* transistor, respectively, and their schematic symbols are shown in figure 8.2.

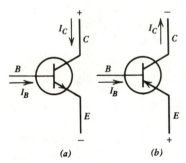

Fig. 8.2 Symbols for bipolar transistors.
(*a*) *npn*. (*b*) *pnp*.

It appears from the construction of the transistor that the emitter and collector ought to be interchangeable. Such is not the case, however, for several reasons. First, the emitter is usually more heavily doped than either the base or collector to minimize the recombination of charge carriers in the base region and increase the amplification. Second, the collector is usually physically larger than the emitter, because most of the voltage drop and hence the heat production occurs at the collector-to-base junction. To effectively dissipate this heat, the collector is often connected to the metal case of the transistor. Finally, the reverse breakdown voltage of the base-to-emitter junction is typically much less than the breakdown voltage of the collector-to-base junction. A transistor will often work if connected backward, but it's not likely to work very well.

8.2 Collector Characteristics

As with the other three-terminal, active devices, the bipolar transistor operation is described by a set of curves called the **collector characteristics**. A typical set of characteristics is shown in figure 8.3. The most proper way to analyze a transistor

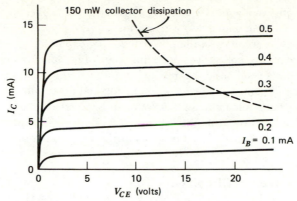

Fig. 8.3 Typical bipolar transistor collector characteristics.

circuit is to follow the same procedure that was used with vacuum tubes and FETs. The operating point is determined by first drawing the load line using the dc open circuit collector-to-emitter voltage and the short-circuit collector current. The intersection of the load line with the collector-current curve corresponding to the appropriate dc base current I_B would then specify the operating point. Then for small variations about the operating point, the ac linear equivalent circuit could be represented by either a Thevenin or Norton equivalent circuit in which the source value and resistance are given by the appropriate partial derivatives.

Fortunately, it happens that a simpler procedure is usually adequate. Inspection of figure 8.3 shows that over most of the range of operation the curves are quite straight, nearly horizontal, and evenly spaced. In fact, it appears that to a good approximation the collector current is simply proportional to the base current independent of operating point:

$$I_C = \beta I_B \qquad (8.1)$$

The proportionality constant, beta (β), is a dimensionless number with a typical value of about 100.

Although the value of β is nearly constant over most of the range of the collector characteristics, at very low values of collector current (near cutoff), the beta varies approximately linearly with collector current. The beta also increases with increasing temperature.

As with the other three-terminal active devices, there is a maximum power that the collector can dissipate without overheating and damaging the transistor. This value can usually be increased substantially by mounting the transistor on a heat sink which conducts heat away from the transistor and dissipates it by convection and radiation. The dashed curve in figure 8.3 shows a collector dissipation of 150 mW, which is a typical limit for a small transistor without a heat sink. The operating point should be chosen so that it lies below such a curve. Transistors, unlike tubes, are very unforgiving when their ratings are exceeded. Because of their small size and mass, even a momentary excursion above the rated collector dissipation will usually cause irreparable damage to a transistor.

The simplest way to increase the power handling capability of a transistor is to increase the size of the collector-to-base junction. The junction capacitance inevitably increases as a result, and the maximum usable frequency is decreased. There is thus an inverse relationship between power rating and maximum frequency. Much of the current effort in semiconductor development is aimed at producing high-power transistors that will operate at high frequencies. As progress is made, vacuum tubes are gradually being replaced with transistors in those few remaining applications.

8.3   Linear Equivalent Circuits

Because of the nearly constant value of beta through most of the operating range of the transitor, it is often possible to use an especially simple model for the transistor that is valid for both ac and dc voltages. Such a model is essentially a Norton equivalent circuit in which the Norton current (βI_B) is dependent on the base current I_B, the Norton resistance is infinite, and the base-to-emitter junction is a short circuit. Such a model, as shown in figure 8.4(a), will define an **ideal transistor**. Note that

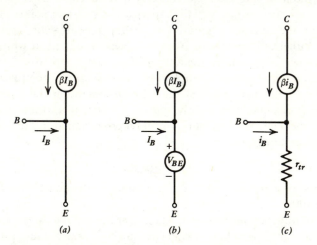

Fig. 8.4   Transistor linear equivalent circuit models. (a) Ideal transistor. (b) dc equivalent circuit of ideal germanium or ideal silicon transistor. (c) ac equivalent circuit of real transistor.

the model is valid only if the voltages and currents all have the proper polarities, and if the transistor is not too close to saturation or cutoff. Such a model will suffice for analyzing most of the circuits encountered in this text.

Unfortunately, the base-to-emitter junction of a transistor is not as good a short circuit as the grid-to-cathode of a vacuum tube is an open circuit. Consequently, a more complicated model is required when accurate results are desired. For calculating the dc voltages and currents, it often suffices to simply add a constant dc voltage of 0.2 V for a germanium transistor or 0.6 V for a silicon transistor in series with the emitter to represent the dc base-to-emitter forward voltage drop, as shown in figure 8.4(b). The voltage source would be positive for an *npn* and negative for a *pnp* transistor. This is the model that will be used to calculate the dc voltages and currents whenever a circuit is said to contain **ideal silicon** or **ideal germanium** transistors, whereas the ac linear equivalent circuit is assumed to be as in figure 8.4(a).

The ac characteristics of a real transistor can be represented to an accuracy that will suffice for all purposes of this text by the ac linear equivalent circuit of figure 8.4(c). The value of β will, in general, depend somewhat on the operating point that is determined using the dc model of figure 8.4(b). The resistance in series with the emitter is called the **transresistance**, and it is a sum of two parts:

$$r_{tr} = r_{\text{ohmic}} + r_d \tag{8.2}$$

The r_{ohmic} term is constant independent of operating point and has a typical value of a few ohms for most transistors. It can usually be neglected. The r_d term is called the **dynamic resistance**, and it can be calculated from the derivative of the V versus I characteristic for a *pn*-junction diode as given by equation 6.3:

$$r_d = \frac{dV_{BE}}{dI_E} = \frac{kT}{eI_E} \simeq \frac{0.026}{I_E} \tag{8.3}$$

The dynamic resistance is an ac resistance given by the slope of the V_{BE} versus I_E curve, and it depends on the dc emitter current at the operating point.

Although the models described above will suffice for the purposes of this text, two additional models are mentioned for the sake of completeness. The first is the **T network equivalent circuit** shown in figure 8.5(a). It closely resembles the model of figure 8.4(c), except that it includes a collector resistance r_C which is determined from the slope of the collector characteristic at the operating point:

$$r_C = \frac{1}{\left(\dfrac{\partial I_c}{\partial V_{CE}}\right)_{I_B = \text{constant}}} \tag{8.4}$$

and a base resistance r_B which is usually on the order of a few ohms. As with any ac linear equivalent circuit, the values of the parameters β, r_C, r_E, and r_B depend on the dc voltages and currents that determine the operating point.

Because the T network equivalent circuit has four parameters, it is complete in the sense that the transistor behavior is exactly predicted, at least in the small amplitude limit, provided the parameters are precisely known. The reason for this is that a transistor is a three-terminal active device, and so in the linear limit it can be completely specified by four parameters, just as a two-terminal, linear, active device can be specified by two parameters (V_T and R_T or I_N and R_N). For example, if v_{CE}, v_{BE}, i_B, and i_C are known, the other voltage can be determined from $v_{CB} = v_{CE} - v_{BE}$ and the other current from $i_E = i_B + i_C$, so that there are only four independent parameters.

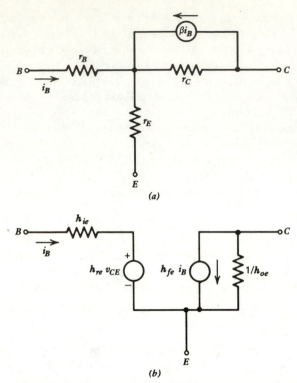

Fig. 8.5 Four-parameter transistor linear equivalent circuit models. (a) T network equivalent circuit. (b) h parameter equivalent circuit.

The most widely used four-parameter transistor model is the **h parameter equivalent circuit** shown in figure 8.5(b). The base-to-emitter junction is represented by a Thevenin equivalent circuit in which the Thevenin voltage depends on v_{CE}. The collector-to-emitter junction is represented by a Norton equivalent circuit in which the Norton current depends on i_B in a manner identical to the simpler models.

Applying Kirchhoff's laws to figure 8.5(b) gives

$$\left.\begin{aligned}
h_{ie} &= \left(\frac{\partial V_{BE}}{\partial I_B}\right)_{V_{CE}=\text{constant}} \\[2mm]
h_{re} &= \left(\frac{\partial V_{BE}}{\partial V_{CE}}\right)_{I_B=\text{constant}} \\[2mm]
h_{fe} &= \left(\frac{\partial I_C}{\partial I_B}\right)_{V_{CE}=\text{constant}} \\[2mm]
h_{oe} &= \left(\frac{\partial I_C}{\partial V_{CE}}\right)_{I_B=\text{constant}}
\end{aligned}\right\} \tag{8.5}$$

The first subscript of the h parameters stands for *input, reverse, forward,* and *output,* respectively. The second subscript (e) denotes the fact that the emitter has been chosen as the terminal that is common to the input and the output. Alternate h parameter representations are sometimes encountered in which another terminal such as the base is common. The resulting h parameters can be written in terms of those in equation 8.5, as follows:

$$\left.\begin{aligned}
h_{ib} &= \frac{h_{ie}}{1 + h_{fe}} \\[1em]
h_{rb} &= \frac{h_{ie}h_{oe}}{1 + h_{fe}} - h_{re} \\[1em]
h_{fb} &= \frac{-h_{fe}}{1 + h_{fe}} \\[1em]
h_{ob} &= \frac{h_{oe}}{1 + h_{fe}} \\[1em]
h_{ic} &= h_{ie} \\[0.5em]
h_{rc} &= 1 - h_{re} \\[0.5em]
h_{fc} &= 1 + h_{fe} \\[0.5em]
h_{oc} &= h_{oe}
\end{aligned}\right\} \quad (8.6)$$

Note that the h parameters have a variety of units (ohms, siemens, dimensionless), and for this reason they are called the **hybrid h parameters**.

By straightforward, although tedious, application of Kirchhoff's laws, the relations between the parameters in the various models can be derived (see problem 8.6). The results are shown in table 8.1, along with typical values for the h

TABLE 8.1 Comparison of Parameters in Various Transistor Models

h Parameter Figure 8.5(b)	T Network Figure 8.5(a)	Real Figure 8.4(c)	Ideal Figure 8.4(a)	Typical Value
h_{ie}	$r_B + \dfrac{r_C r_E (\beta + 1)}{r_C + r_E}$	$(\beta + 1)r_{tr}$	0	$10^3 \ \Omega$
h_{re}	$\dfrac{r_E}{r_C + r_E}$	0	0	10^{-3}
h_{fe}	$\dfrac{\beta r_C - r_E}{r_C + r_E}$	β	β	100
h_{oe}	$\dfrac{1}{r_C + r_E}$	0	0	$10^{-4} \ \mho$

parameters. Note that an ideal transistor is one in which all the h parameters are zero except of h_{fe} which is the same as β. The representation we will use for a real transistor has $h_{fe} = \beta$, $h_{ie} = (\beta + 1)\, r_{tr}$, and the other h parameters equal to zero.

It should be emphasized that all these transistor circuit representations are valid only when the ac voltages and currents are sufficiently small, only when the dc voltages and currents have the correct sign, and only when the transistor is not saturated or cut off.

8.4 Common Emitter Amplifier

The basic bipolar transistor amplifier circuit is the **common emitter amplifier** shown in figure 8.6. It is analogous to the common cathode vacuum tube circuit and the common source FET circuit. The name comes from the fact that the emitter is

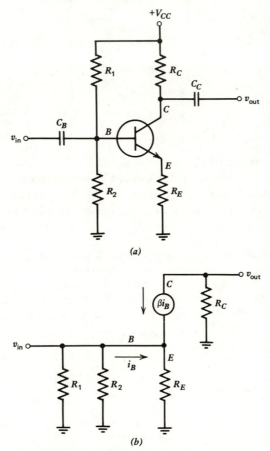

Fig. 8.6 The common emitter amplifier in (a) can be analyzed using the linear equivalent circuit in (b).

common to both the input and the output circuits. We will first analyze the behavior of the circuit, assuming the transistor is ideal. By setting the dc source (V_{CC}) equal to zero, replacing the capacitors with short circuits, and using the linear equivalent circuit for the ideal transistor [figure 8.4(a)], the circuit in figure 8.6(b) is obtained. Applying Ohm's law to R_E gives

$$v_{in} = (\beta i_B + i_B) R_E$$

Similarly, for R_C, Ohm's law gives

$$v_{out} = -\beta i_B R_C$$

The amplification A is given by

$$A = \frac{v_{out}}{v_{in}} = -\frac{\beta i_B}{(\beta + 1) i_B} \frac{R_C}{R_E} \simeq -\frac{R_C}{R_E} \tag{8.7}$$

where the last approximation is valid for $\beta \gg 1$. Note that the amplification is independent of β (for β large), which is fortunate, since the β of transistors of the same type often varies by a factor of two or more, and it would otherwise be difficult to mass-produce amplifiers with specific characteristics. The fact that A is independent of β also means that the transistor behaves in a linear fashion, even for large excursions from its design operating point. Note also that A is negative, as was the case for the common cathode and common source amplifiers.

The output resistance of the common emitter amplifier is determined by dividing the open circuit output voltage $-\beta i_B R_C$ by the short-circuit output current $-\beta i_B$, giving the simple result:

$$R_{out} = R_C \tag{8.8}$$

The input resistance is a little more complicated. It is tempting to set the current source equal to zero and to say that the input resistance consists of R_1, R_2, and R_E in parallel. However, this is not allowed, because the current source has a value proportional to i_B, and a measurement of the resistance requires one to apply a voltage v_{in} which produces a current i_B, so that

$$r_{in} = \frac{v_{in}}{i_B} = \frac{(\beta + 1) i_B R_E}{i_B} \simeq \beta R_E \tag{8.9}$$

The input resistance r_{in} between the transistor base and ground, neglecting R_1 and R_2, is thus not R_E but βR_E (for $\beta \gg 1$). The total input resistance R_{in} is then given by

$$\frac{1}{R_{in}} = \frac{1}{R_1} + \frac{1}{R_2} + \frac{1}{\beta R_E} \tag{8.10}$$

Normally, one chooses R_1 and R_2 sufficiently small that βR_E can be neglected. Then the input resistance is not affected by variations of β, and so it remains nearly constant for transistors with widely varying characteristics and for different operating points for a given transistor.

The reader may wonder why a voltage divider (R_1 and R_2) is used to establish the operating point rather than simply omitting R_2 and obtaining the required dc base current from R_1 alone. The reason is that the variation in beta from one transistor to the next would cause a wide variation of operating points for otherwise identical circuits. If the collector current at the operating point were chosen to be one-half the short-circuit current for a given transistor and then a transistor with a beta twice as great were substituted in the circuit, the circuit would be saturated. Furthermore, such a circuit tends to be thermally unstable. If the transistor heats up, the beta increases, and the collector current rises for a constant base current. This can cause the transistor to heat up even more, further increasing the beta. In an extreme case, **thermal runaway** occurs, the maximum collector dissipation is exceeded, and the transistor is destroyed.

These difficulties are avoided by a proper choice of R_1, R_2, and R_E. R_1 and R_2 establish the dc base voltage and hence the emitter voltage (since $V_E \simeq V_B$). For a given emitter voltage R_E determines the dc emitter current, and hence the collector current (since $I_C \simeq I_E$) and operating point. The circuit characteristics are thus almost entirely independent of the transistor characteristics.

In designing a transistor amplifier, the resistors are chosen as follows:

1. R_C is chosen to provide the desired output resistance ($R_C = R_{\text{out}}$).

2. R_E is then chosen to provide the desired amplification ($R_E = -R_C/A$).

3. R_1 and R_2 are chosen so that their parallel combination is small (say, 10%) compared with βR_E and such that the operating point is at the desired place, usually near the center of the collector characteristics. For example, one normally takes the collector current to be about half the short-circuit ($V_{CE} = 0$) current:

$$I_C \simeq \frac{V_{CC}}{2(R_C + R_E)}$$

Since $I_E \simeq I_C$, the emitter voltage desired is

$$V_E \simeq I_C R_E \simeq \frac{R_E V_{CC}}{2(R_C + R_E)}$$

and so by the voltage divider relation, since $V_B \simeq V_E$,

$$\frac{R_2 V_{CC}}{R_1 + R_2} \simeq \frac{R_E V_{CC}}{2(R_C + R_E)}$$

or

$$\frac{R_1}{R_2} \simeq \frac{2R_C}{R_E} + 1 \tag{8.11}$$

It appears from the above that the amplification $|A|$ can be made arbitrarily large by making R_E small. If R_E is too small, however, the ideal transistor linear equivalent circuit is no longer adequate, and one must consider the transresistance, which appears in series with R_E. In such a case the amplification is

$$A = -\frac{R_C}{(R_E + r_{tr})} \tag{8.12}$$

which depends on the operating point and has a limiting value of $-R_C/r_{tr}$ when $R_E = 0$. The input resistance is also lowered significantly by taking $R_E = 0$. One often connects an **emitter bypass capacitor** in parallel with R_E in a manner analogous to figures 7.7(a) and 7.17(a). The amplification in such a case still depends on the transresistance and hence on the operating point, but at least the operating point is determined by the external resistors rather than by the characteristics of the transistor itself. The use of the external resistor R_E to reduce the amplification and mask the inherent nonlinearity of the transistor is an example of **negative feedback** which will be discussed in some detail in the next chapter.

8.5 Emitter Follower Circuit

The bipolar transistor can be used in a circuit analogous to the cathode follower discussed in section 7.4. Such a circuit as shown in figure 8.7(a) is called an **emitter follower** or a **common collector amplifier**. Assuming the transistor to be ideal leads to the ac linear equivalent circuit in figure 8.7(b). It is readily apparent that $v_{out} = v_{in}$. With a more realistic transistor model, a resistance r_{tr} would be in series with the emitter, as shown in figure 8.6(c), and the output voltage would be given by the voltage divider relation:

$$v_{out} = \frac{v_{in}R_E}{R_E + r_{tr}}$$

so that the amplification is

$$A = \frac{R_E}{R_E + r_{tr}} \tag{8.13}$$

The input resistance is the same as for the common collector amplifier:

$$\frac{1}{R_{in}} = \frac{1}{R_1} + \frac{1}{R_2} + \frac{1}{\beta R_E} \tag{8.14}$$

The output resistance is zero if v_{in} is connected to an ideal voltage source (no internal resistance) and the transistor is ideal as shown in figure 8.7(b). For the more realistic transistor model of figure 8.7(c) the output resistance is

$$R_{out} = \frac{r_{tr}R_E}{r_{tr} + R_E} \tag{8.15}$$

as can be seen by examining figure 8.7(c) with $v_{in} = 0$. In contrast to the vacuum tube and FET, the output resistance of the emitter follower depends on the internal resistance of the source connected to the input, and the input resistance also depends

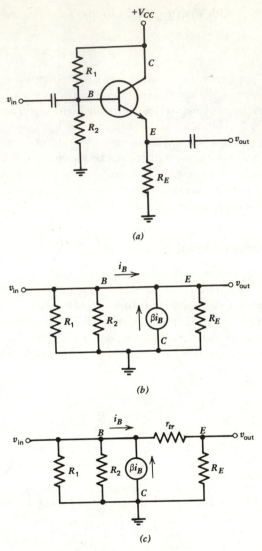

Fig. 8.7 The emitter follower circuit in (*a*) can be analyzed using the linear equivalent circuit in (*b*) if the transistor is ideal, or using the more realistic model in (*c*) if more accurate results are required.

on the resistance of the load connected to the output. The output resistance is lowest when the source resistance is low, and the input resistance is lowest when the load resistance is low. The emitter follower is thus a near-unity voltage gain impedance transformer, but it does not isolate the input from the output as thoroughly as does the cathode follower or source follower.

8.6 Common Base Amplifier

As a final example of a single-transistor linear amplifier circuit, we will consider the **common base amplifier** shown in figure 8.8(a). Treating the capacitors as short circuits for ac and the transistor as real leads to the linear equivalent circuit in figure 8.8(b). The transresistance has to be included in this circuit, because otherwise it would be impossible to have a voltage between the input and ground.

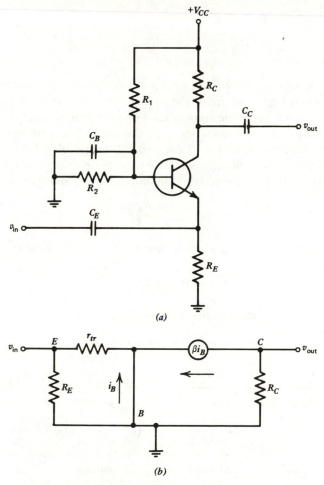

(a)

(b)

Fig. 8.8 The common base circuit in (a) can be analyzed using the linear equivalent circuit in (b).

The base current i_B in figure 8.8(b) from Kirchhoff's current law is a sum of two parts, both negative:

$$i_B = -\frac{v_{in}}{r_{tr}} - \beta i_B$$

Solving for i_B gives

$$i_B = -\frac{v_{in}}{(\beta + 1)r_{tr}}$$

The output voltage is

$$v_{out} = -\beta i_B R_C = \frac{\beta v_{in} R_C}{(\beta + 1)r_{tr}}$$

Therefore the amplification is

$$A = \frac{v_{out}}{v_{in}} = \frac{\beta R_C}{(\beta + 1)r_{tr}} \simeq \frac{R_C}{r_{tr}} \tag{8.16}$$

where the last approximation is valid for $\beta \gg 1$. Like the common emitter amplifier, the amplification is large, but unlike the common emitter circuit, the output is not inverted. The dependence of the amplification on the transresistance can be reduced either by placing a resistor in series with the input (see problem 8.14) or by eliminating the capacitor from the base to ground (see problem 8.15).

The input resistance of the common base amplifier, as can be seen by inspection of figure 8.8(b), is given by the parallel combination of r_{tr} and R_E:

$$R_{in} = \frac{r_{tr} R_E}{r_{tr} + R_E} \tag{8.17}$$

and hence is quite small, since r_{tr} is usually small. The output resistance is the same as for the common emitter circuit:

$$R_{out} = R_C \tag{8.18}$$

TABLE 8.2 Characteristics of the Three Types of Transistor Amplifiers

	Common Emitter	Emitter Follower	Common Base
Amplification (voltage) $A =$	Medium $-\dfrac{R_C}{R_E + r_{tr}}$	Small $\dfrac{R_E}{R_E + r_{tr}}$	Large $\dfrac{R_C}{r_{tr}}$
Input resistance $R_{in} =$	Medium $\left(\dfrac{1}{R_1} + \dfrac{1}{R_2} + \dfrac{1}{\beta R_E}\right)^{-1}$	Medium $\left(\dfrac{1}{R_1} + \dfrac{1}{R_2} + \dfrac{1}{\beta R_E}\right)^{-1}$	Small $\dfrac{r_{tr} R_E}{r_{tr} + R_E}$
Output resistance $R_{out} =$	Medium R_C	Small $\dfrac{r_{tr} R_E}{r_{tr} + R_E}$	Medium R_C

In summary, table 8.2 shows the characteristics of the three basic types of transistor amplifier circuits. One should note the similarity to the three vacuum tube circuits listed in table 7.1.

8.7 Transistor Voltage Regulators

The amplifier circuits previously discussed account for only a small fraction of the possible applications of the bipolar transistor. In this section we will consider how transistors can be used to maintain a constant output voltage across a load in which the current may vary drastically. Such a regulator is often used with a rectifier and filter circuit in a device called a **regulated power supply** which behaves much like an ideal dc voltage source. We will consider two types of regulators, the **series regulator** and the **parallel** (or **shunt**) **regulator**.

Figure 8.9(a) shows the basic series regulator. Its operation is very easy to

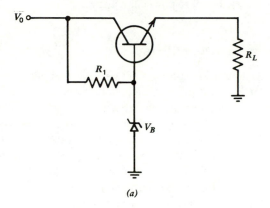

(a)

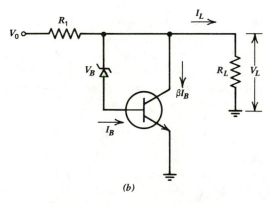

(b)

Fig. 8.9 Transistor voltage regulators. (a) Series. (b) Parallel.

explain. The Zener diode holds the base at a voltage V_B over a wide range of conditions. The circuit is essentially an emitter follower, and so the emitter (and hence the load) voltage is the same as (or a constant 0.6 V for a silicon transistor less than) the base voltage. The advantage of the series regulator over the simple Zener diode regulator discussed in section 6.7 is that the power dissipated by the regulator is considerably smaller, especially in the no-load ($R_L = \infty$) condition.

The basic shunt regulator is shown in figure 8.9(b). Whenever the output voltage V_L tries to rise above V_B, neglecting the small base-to-emitter voltage, the Zener diode conducts, and a large current I_B flows into the base of the transistor. This current is amplified by a factor of β and increases the voltage drop across R_1 until V_L drops to V_B. The transistor amplifies the effect of the Zener diode so that the Zener need not dissipate appreciable power. The transistor does, however, dissipate as much power as would a Zener diode at the same place in the circuit. The only real advantage is that high-power transistors are usually less expensive than the equivalent high-power Zener diode. The shunt regulator dissipates the most power for small I_L, whereas the series regulator dissipates the most power for large I_L.

8.8   Multiple-Transistor Amplifiers

Depending on the configuration, the voltage gain of a single transistor amplifier is limited to approximately $-R_C/r_{tr}$. To achieve higher gains, to improve stability, and to increase bandwidth, amplifiers usually employ several stages of amplification. A straightforward approach is to use two or more single transistor amplifiers as previously described, with the output of one connected to the input of the next, and so on. In such a case it is tempting to calculate the overall amplification by simply multiplying together the amplification of the various stages. This would be correct, however, only if the input resistance of each stage is very large compared with the output resistance of the previous stage. In fact, for maximum power transfer one generally designs each stage so that its input resistance is approximately equal to the output resistance of the previous stage. When this is done, the amplification of each stage is reduced to half the value it would have with no output load, and the overall amplification is reduced by a factor of 2^n, where n is the total number of stages. This reduction of amplification is called **loading**.

An alternate configuration is the **Darlington pair** shown in figure 8.10(a), which can be analyzed using the linear equivalent circuit in figure 8.10(b). Such a circuit is identical to a single transistor with a β given by

$$\beta = (\beta_1 + 1)\beta_2 \tag{8.19}$$

which can easily exceed 10^4. The total dc base-to-emitter voltage drop is the sum of the base-to-emitter drops for each transistor (i.e., 1.2 V for silicon). Such a circuit is often made in a single package with three terminals so as to behave like a single, very high beta transistor.

A practical difficulty with the Darlington configuration is that the transistor

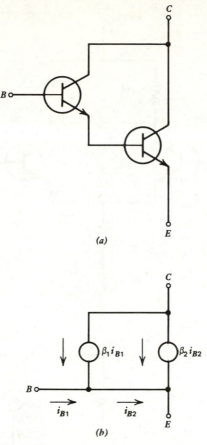

(a)

(b)

Fig. 8.10 The Darlington pair in (*a*) can be analyzed using the linear equivalent circuit in (*b*).

types have to be chosen very carefully if the full benefit of the high beta is to be obtained. To avoid saturating the second transistor, its base current, and hence the collector current of the first transistor, has to be very small. An input transistor thus has to be chosen that has a high beta at low values of collector current.

Two transistors can be connected as in figure 8.11(*a*) to form a **difference amplifier** which can be analyzed using the linear equivalent circuit in figure 8.11(*b*). The transistors are assumed to be identical, and it is necessary to consider the transresistance so as to allow a voltage difference between the two bases. For $R_E \gg r_{tr}$, the current i_1 produced by v_1 and v_2 can be determined, using the superposition theorem (alternately, connect v_2 and then v_1 to ground):

$$i_1 \simeq \frac{v_1}{2\beta r_{tr}} - \frac{v_2}{2\beta r_{tr}}$$

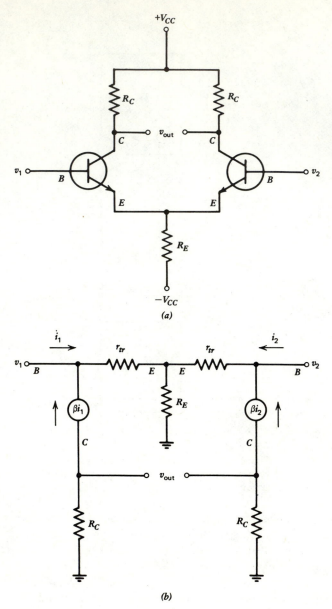

Fig. 8.11 The difference amplifier in (a) can be analyzed using the linear equivalent circuit in (b).

Similarly, i_2 is given by

$$i_2 \simeq \frac{v_2}{2\beta r_{tr}} - \frac{v_1}{2\beta r_{tr}}$$

The voltage drop v_{out} is

$$v_{\text{out}} = \beta i_1 R_C - \beta i_2 R_C \simeq \frac{R_C}{r_{tr}}(v_1 - v_2) \qquad (8.20)$$

Such a circuit is useful for subtracting two voltages and amplifying the difference. This is extremely useful whenever it is necessary to measure the voltage between two points in a circuit, neither of which is grounded, and to reference the measured voltage to ground. It is also useful for amplifying dc voltages, since no capacitors are used. A characteristic of most dc amplifiers is that both a positive and negative power supply voltage are required.

The quality of a difference amplifier is expressed in terms of its **common mode rejection ratio** (CMRR). The CMRR is the ratio of the voltage that must be applied at the two inputs in parallel (v_1 and v_2) to the difference voltage ($v_1 - v_2$), for the output to be of the same magnitude. The CMRR of the difference amplifier in figure 8.11 is theoretically infinite. If any of the corresponding components are not identical, a finite CMRR will result (see problem 8.19). Difference amplifiers usually have some means of adjusting for small asymmetries in the circuit to maximize the CMRR. It is relatively easy to obtain a CMRR of $\sim 10^3$ to 10^4 over a narrow range of frequencies, but much more difficult when the amplifier bandwidth is large.

Another useful circuit is the **complementary-symmetry amplifier** shown in figure 8.12. It uses an *npn* and a *pnp* transistor of otherwise identical characteristics.

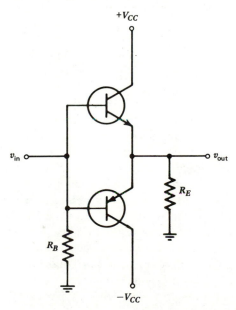

Fig. 8.12 A complementary-symmetry amplifier using an *npn* and a *pnp* transistor biased to cutoff (class B).

For $v_{in} = 0$, both transistors are biased to cutoff (i.e., $I_B = 0$). For $v_{in} > 0$, the upper transistor conducts and behaves like an emitter follower while the lower transistor is cut off. For $v_{in} < 0$, the lower transistor conducts while the upper transistor is cut off. In addition to being useful as a dc amplifier, such a circuit conserves power, because the operating point for both transistors is near $I_C = 0$. Of course, the characteristics of a transistor (β, r_{tr}) are not very constant near $I_C = 0$, and so the circuit is not very linear for small signals or near the zero-crossing point of a large signal.

Since it is difficult to obtain *npn* and *pnp* transistors that are accurately matched, it is more common to find circuits that use two transistors of the same type in what is called a **push-pull amplifier circuit** as shown in figure 8.13. In this case, the

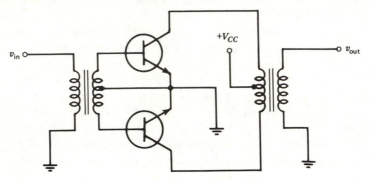

Fig. 8.13 A push-pull amplifier uses two matched transistors biased to cutoff (class B).

transistors are connected as common emitter amplifiers, although an emitter follower and common base configuration are also possible. The push-pull amplifier resembles the full-wave rectifier in its use of center-tapped transformers. The use of transformers in a low-frequency circuit of this type is undesirable in terms of cost, weight, space, and linearity, but a transformer does provide considerable flexibility, in that it enables the designer to match input and output resistances in a way that optimizes the overall performance of the circuit. As with the complementary symmetry amplifier, the push-pull amplifier is normally operated with $I_B = 0$, so that neither transistor dissipates power until an ac input signal is applied.

A circuit biased in such a fashion is called a **class B amplifier**, in contrast with the **class A amplifiers** previously discussed, in which the operating point is near the center of the collector characteristics. Circuits are sometimes constructed in which the base is reverse-biased so as to conduct over only a small fraction of the period of the input signal. Such circuits are called **class C amplifiers** and are used primarily for amplifying high-frequency signals having a narrow Fourier spectrum. The input and output voltages for the three types of amplifiers are shown in figure 8.14.

Although class C amplifiers are the most efficient of the three, they produce drastic distortion of the input signal and so are normally used with high Q resonant

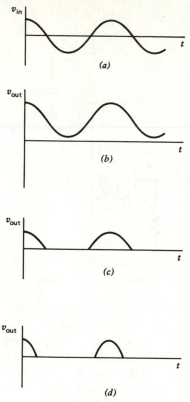

Fig. 8.14 Voltage waveforms for various classes of amplifiers. (*a*) Input voltage. (*b*) Class A amplifier. (*c*) Class B amplifier. (*d*) Class C amplifier.

circuits in their output to attenuate the unwanted harmonics, as shown in figure 8.15. Alternately, by tuning the output circuit to one of the harmonic frequencies, a class C amplifier can be used as a frequency multiplier. The resonant *LC* circuit in the collector is called the **tank circuit**, and it is tuned to the frequency of the input voltage or one of its harmonics to produce a nearly sinusoidal output despite the highly nonlinear nature of the class C amplifier.

Figure 8.15 also shows how simply the proper base bias for class C operation can be obtained. Use is made of the fact that the base-to-emitter junction is a diode rectifier, and so the capacitor C_B will tend to charge up with a dc voltage that keeps the base reverse-biased during most of the cycle of the input waveform.

The high efficiency of the class C amplifier results from the fact that the transistor behaves much like a switch. Most of the time it is cut off and hence draws no current. When it does conduct, it conducts strongly so that the collector-to-emitter voltage is small. In either case the power dissipated by the transistor is small. Note that the

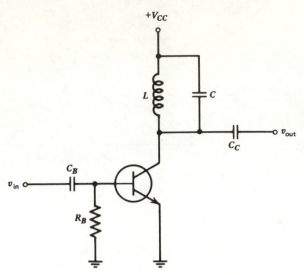

Fig. 8.15 Class C amplifier with resonant output circuit.

class C amplifier is operated far from its linear regime. Consequently, the linear equivalent circuits presented earlier are of virtually no use in analyzing such a circuit.

When good linearity over a wide range of frequencies is required, as in a high-fidelity audio amplifier, a class A amplifier must be used, with some sacrifice of efficiency. For any amplifier, the **efficiency** η is defined as the ac power delivered to the load divided by the total power produced by all the sources. Typical efficiencies are ~ 10–30% for a class A amplifier and ~ 70–80% for a class C amplifier.

8.9 Summary

The bipolar transistor operates in a manner analogous to the vacuum tube and FET except that it is controlled by a current rather than by a voltage. It is inherently a low-input resistance device, in contrast to the vacuum tube and FET. A bipolar transistor tends to be more linear than the other devices, and it usually suffices to neglect the collector resistance and to ignore the variation of β with operating point. The base-to-emitter junction is a forward-biased diode, and so it has a small, nearly constant, dc voltage drop which must sometimes be considered. In addition, the emitter behaves as if it has an ac internal series resistance r_{tr} that depends on the dc emitter current.

The transistor can be used as an amplifier in either the common emitter, common collector (emitter follower), or common base configuration. The common emitter circuit has a large amplification but a high output resistance. The emitter follower has an amplification slightly less than one, but a very low output resistance. The common base circuit has a large amplification and a very low input resistance.

The transistor can also be used as a voltage regulator either in series or parallel with the load. With two or more transistors, the variety of possible circuits is very large. The Darlington pair, the difference amplifier, the complementary-symmetry amplifier, and the push-pull amplifier are four common examples. The way in which an amplifier is biased allows one to trade off linearity for efficiency. Class A amplifiers are the most linear, and class C amplifiers are the most efficient.

Problems

8.1   Using the collector characteristics in figure 8.3, determine the operating point for the circuit below:

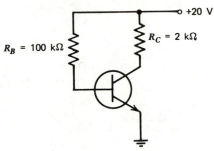

8.2   Calculate the value of r_C for the problem above.

8.3   Calculate the value of V_{CE} and I_C for the circuit in problem 8.1 if R_B is changed to 10 kΩ.

8.4   Calculate the value of V_{BE} and r_{tr} in the circuit in problem 8.1 assuming the transistor is germanium with $r_{ohmic} = 4\ \Omega$.

8.5   Show that if $r_C = \infty$ in the T network model of figure 8.5(a), the input and output currents and voltages are the same as for the real transistor model of figure 8.4(c), and derive an expression for r_{tr} in terms of β, r_B, and r_E.

8.6   By application of Kirchhoff's laws, derive the h parameters for the T network given in table 8.1.

8.7   Assume figure 8.6 contains an ideal silicon transistor with $\beta = 100$ and $V_{CC} = 10$ V, $R_1 = 20$ kΩ, $R_2 = 5$ kΩ, $R_E = 1$ kΩ, and $R_C = 5$ kΩ. Calculate V_B, V_E, V_C, I_B, I_E, and I_C.

8.8   For the circuit described in problem 8.7, calculate R_{in}, R_{out}, and A. How would these values be changed if an emitter bypass capacitor were added to the circuit? Assume $r_{ohmic} = 1.4\ \Omega$.

8.9   For the circuit described in problem 8.7, calculate the amplification A assuming the output is connected to a resistor $R_L = 7.5$ kΩ.

8.10   Design a common emitter amplifier with $R_{out} = 5$ kΩ and $A = -10$ using a 15-V power supply and an ideal transistor with $\beta = 100$. Calculate its input resistance.

8.11   For the circuit below, calculate the dc and the ac parts of the output voltage $V_{out}(t)$, assuming the transistor is ideal with $\beta = 99$.

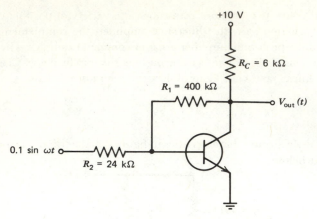

8.12 The circuit below contains a real germanium transistor with $\beta = 50$ and $r_{tr} = 2.0\ \Omega$. Calculate the input resistance R_{in} and the voltage v_L across the load resistor R_L.

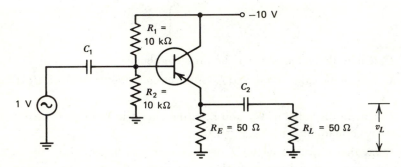

8.13 For the circuit in problem 8.12, estimate the values required for C_1 and C_2 such that the 3-dB point will occur at $f_c = 25$ Hz.

8.14 Calculate the input resistance and amplification of the circuit on the following page, assuming the transistor is ideal and the capacitors are short circuits for ac.

8.15 Calculate the input resistance and amplification of the circuit in figure 8.8(*a*) with the capacitor C_B between base and ground removed, using $R_1 = 40$ kΩ, $R_2 = 10$ kΩ, $R_C = 4$ kΩ, and $R_E = 1200\ \Omega$, assuming the transistor is ideal with $\beta = 99$ and the other capacitors are short circuits for ac.

8.16 The circuit on the following page acts as a constant current source. Show that the current I is independent of R_L for R_L below a critical value, and calculate that value of R_L. Assume that the transistor is ideal.

8.17 Assume the circuit in figure 8.9(*a*) contains an ideal transistor with $\beta = 39$. Calculate the maximum current in R_L for which the circuit regulates properly, assuming $V_0 = 20$ V, $V_B = 10$ V, and $R_1 = 400\ \Omega$. Calculate the power dissipated in the load, in the transistor, in resistor R_1, and in the Zener diode under the above conditions.

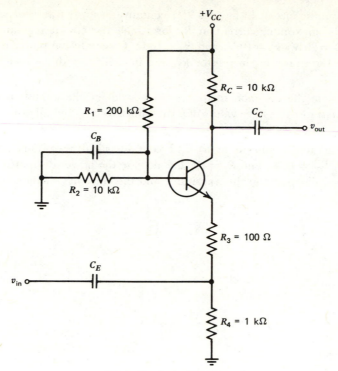

Illustration for problem 8.14.

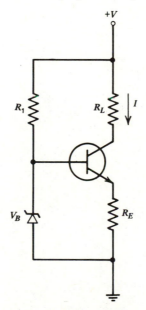

Illustration for problem 8.16.

8.18   Assume the circuit in figure 8.9(*b*) contains an ideal transistor with $\beta = 39$. Calculate the maximum current in R_L for which the circuit regulates properly, assuming $V_0 = 20$ V, $V_B = 10$ V, and $R_1 = 10\ \Omega$. Calculate the power dissipated in the load, in the transistor, in resistor R_1, and in the Zener diode under the above conditions.

8.19   Calculate the common mode rejection ratio for the circuit in figure 8.11, assuming the transistors have values of β that differ by 1% and all other parameters are identical.

8.20   Assume the circuit in figure 8.12 contains ideal transistors and that $v_{in} = 10 \sin \omega t$, $V_{CC} = 10$ V, and $R_E = 10\ \Omega$. Calculate the power dissipated in the load R_E, the power dissipated in the transistors, and the efficiency of the amplifier.

chapter 9

Operational Amplifiers

9.1 Operational Amplifier Characteristics

A high-gain, multistage, dc amplifier containing many individual transistors and resistors can be considered as a single, active, circuit component called an **operational amplifier** (**op amp** for short). Such a circuit is usually miniaturized and fabricated on a single chip of silicon in what is called an **integrated circuit** (**IC**). Such integrated circuits often contain hundreds of individual components, and when mass produced, are comparable in size and cost to a single transistor. The op amp is such a useful device that it has become the basic building block of analog electronics and has revolutionized the way in which complicated electronic circuits are designed and constructed.

Figure 9.1 shows a schematic diagram of a typical, low-cost, general-purpose, operational amplifier. It is not necessary to understand its operation in detail. However, one should notice that the input stage is a difference amplifier, the output stage is a complementary symmetry amplifier, and all stages are dc coupled. These amplifier circuits were described in section 8.8. This chapter will describe some of the properties and uses of operational amplifiers.

In practice, it is not necessary to know what is contained within an op amp in order to use it. Its behavior is completely specified by the relations of the voltages and currents at its terminals in the same way that other devices such as the vacuum tube and transistor are specified by the V-I relations at their terminals. The op amp is basically a four-terminal device, with two inputs, one output, and a common terminal, which is usually connected to ground. In addition, a plus and minus dc voltage must be supplied, and extra terminals are often provided to compensate for certain nonideal properties of the device, such as frequency response and dc offset. Often the ground terminal is omitted from the op amp, and the output voltage is referenced instead to the midpoint of the positive and negative dc supply voltages as determined from a voltage divider internal to the op amp.

The most important characteristics of the op amp are the **open-loop voltage gain**, A_0, which is a function of input voltage, frequency, the input resistance r_{in}, and the output resistance r_{out}. A typical plot of output voltage versus input voltage difference is shown in figure 9.2. Note that the device is nonlinear, as would be

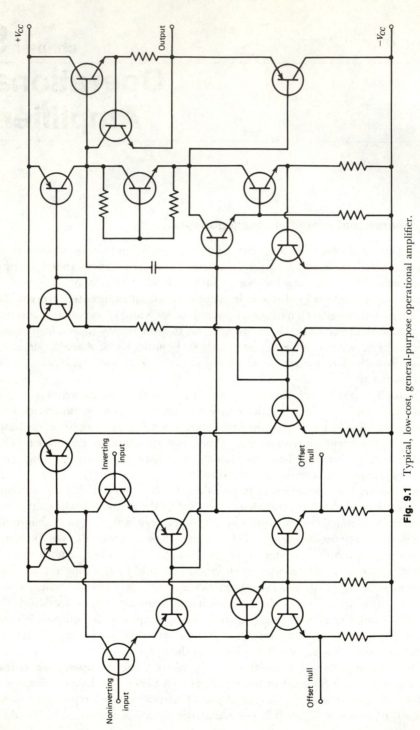

Fig. 9.1 Typical, low-cost, general-purpose operational amplifier.

196

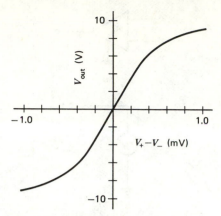

Fig. 9.2 Output versus input voltage for a typical real operational amplifier.

expected, since it contains nonlinear components. In particular, the output voltage always saturates at a value close to but slightly below the dc power supply voltage. The device shown has an open-loop voltage gain of $A_0 \sim 10^4$, since the output reaches ~ 10 V when the input voltage difference is ~ 1 mV.

The open-loop voltage gain of op amps in common use is typically in the range of 10^2 to 10^6 (40 to 120 dB). Most op amps have input resistances in the range of 10^5 to 10^7 Ω. Special op amps with MOSFET input amplifiers have r_{in} of 10^{11} Ω or higher. The output resistance of op amps is usually in the range of about 10 to 1000 Ω.

As with the other components, it is useful to define an ideal op amp as one in which A_0 is constant, r_{in} is infinite, and r_{out} is zero. The symbol for an ideal op amp is shown in figure 9.3(a). Whenever the symbol A_0 is omitted, for the purposes of this text, it is assumed to be infinite. The ideal op amp thus behaves like an ideal voltage source with

$$V_{out} = A_0(V_+ - V_-) \tag{9.1}$$

as shown in figure 9.3(b). A better representation for a real op amp is the Thevenin equivalent circuit shown in figure 9.3(c). The inputs V_+ and V_- are called the **noninverting** and the **inverting inputs**, respectively. Like the other active devices studied, the op amp contains a dependent source, that is, a source whose value depends on the value of a voltage elsewhere in the circuit.

9.2 Negative Feedback

Operational amplifiers are usually used in circuits that provide **negative feedback**. One example of negative feedback has already been encountered in the common emitter amplifier in figure 8.6, where part of the output voltage appears across the resistor R_E causing the base-to-emitter voltage v_{BE} to be reduced to a very low value relative to v_{in}. The result of this negative feedback is to reduce the overall

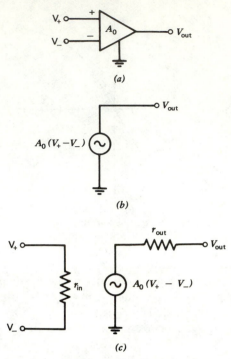

Fig. 9.3 (a) Symbol for an ideal op amp. (b) Linear equivalent circuit. (c) Representation of a real op amp.

amplification of the circuit but to make it insensitive to the beta, nonlinearities, and other nonideal characteristics of the transistor.

Negative feedback is extremely useful. Consider an arbitrary amplifier circuit in which the amplification in the absence of feedback is A_0. If a fraction f of the output is returned and subtracted from the input V_{in}, then the output is given by

$$V_{out} = A_0(V_{in} - fV_{out})$$

In the presence of feedback, the amplification is then given by

$$A = \frac{V_{out}}{V_{in}} = \frac{A_0}{1 + A_0 f}$$

If the product $A_0 f$ is sufficiently large ($\gg 1$), the amplification becomes simply $A = 1/f$, independent of A_0. Thus the fact that A_0 is not really a constant for most nonlinear circuits is of little consequence in the behavior of the circuits. Even though A_0 may change drastically with input signal amplitude, frequency, power supply voltage, temperature, age, and so on, the circuit operation is determined only by the fraction f of negative feedback.

Figure 9.4 shows two common forms of operational amplifier negative feedback.

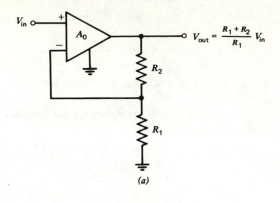

$$V_{out} = \frac{R_1 + R_2}{R_1} V_{in}$$

(a)

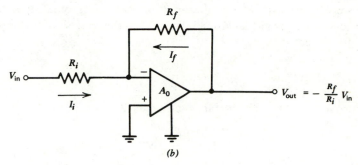

$$V_{out} = -\frac{R_f}{R_i} V_{in}$$

(b)

Fig. 9.4 Examples of op amp feedback. (*a*) Voltage feedback. (*b*) Operational feedback.

These circuits are easily analyzed in the limit $A_0 = \infty$. For such a case, the output voltage can be finite only if $V_+ = V_-$. One of the most perplexing properties of the ideal op amp with negative feedback and infinite A_0 is that the input behaves simultaneously like an open and a short circuit. It behaves like an open circuit, because the input resistance is very high, and hence no current flows into either input terminal. It behaves like a short circuit, because the voltage difference between the two input terminals is very small. The input is thus unlike any circuit element previously encountered. It is, in fact, simpler, once one gets used to its unusual properties.

The circuit in figure 9.4(*a*) has what is called **voltage feedback**, and it uses a voltage divider to supply a fixed fraction of the output at the inverting input terminal:

$$V_{in} = V_+ = V_- = \frac{R_1 V_{out}}{R_1 + R_2} \tag{9.2}$$

or

$$V_{out} = \frac{R_1 + R_2}{R_1} V_{in} \tag{9.3}$$

The amplification is then given by the ratio of the resistors:

$$A = \frac{V_{out}}{V_{in}} = 1 + \frac{R_2}{R_1} \qquad (9.4)$$

independent of the properties of the op amp in a manner reminiscent of the common emitter transistor amplifier.

For the circuit in figure 9.4(b), which has **operational feedback**, the current in R_i is

$$I_i = \frac{V_{in}}{R_i}$$

since $V_- = V_+ = 0$. In such a case, the inverting input of the op amp is called a **virtual ground**, since it is always at the same voltage as the grounded, noninverting input. The concept of the virtual ground is central to the analysis of op amp circuits. Similarly, the current in R_f is

$$I_f = \frac{V_{out}}{R_f}$$

Since no current can flow into the input terminals, I_f is equal to $-I_i$, or

$$V_{out} = \frac{R_f}{R_i} V_{in} \qquad (9.5)$$

and the amplification is

$$A = \frac{V_{out}}{V_{in}} = -\frac{R_f}{R_i} \qquad (9.6)$$

The amplification is determined only by the ratio of two resistors, but in this case the output is inverted (shifted in phase by 180°).

Equations 9.4 and 9.6 imply that an arbitrarily large amplification can be obtained by an appropriate choice of resistors. Such is, of course, not the case. If R_2 or R_f is made very large, the feedback is eliminated, and the amplification approaches A_0. The equivalent circuit corresponding to figure 9.4(a) for A_0 finite is shown in figure 9.5(a). From the voltage divider relation,

$$V_- = \frac{R_1}{R_1 + R_2} V_{out}$$

Since the output voltage is

$$V_{out} = A_0 (V_{in} - V_-)$$

the V_- can be eliminated from the above equation (provided $A_0 \gg 1$) to give

$$V_{out} \simeq \frac{A_0(R_1 + R_2)}{A_0 R_1 + R_2} V_{in}$$

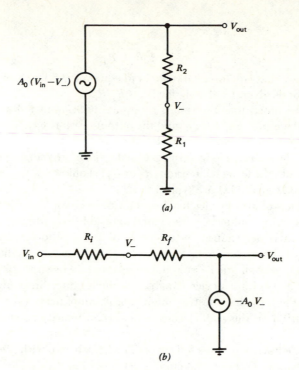

Fig. 9.5 Equivalent circuits for calculating the amplification of the circuits in figure 9.4 for A_0 finite. (*a*) Voltage feedback. (*b*) Operational feedback.

or

$$A = \frac{V_{\text{out}}}{V_{\text{in}}} \simeq \frac{A_0(R_1 + R_2)}{A_0 R_1 + R_2} \qquad (9.7)$$

One should note that equation 9.7 reduces to equation 9.4 for $A_0 \gg R_2/R_1$ and that the amplification is given by $A = A_0$ for $A_0 \ll R_2/R_1$.

In a similar fashion the limiting amplification of the circuit in figure 9.4(*b*) can be calculated, using the equivalent circuit shown in figure 9.5(*b*). Equating currents in the two resistors gives

$$\frac{V_{\text{in}} - V_-}{R_i} = \frac{V_- - V_{\text{out}}}{R_f}$$

also

$$V_{\text{out}} = -A_0 V_-$$

Combining the above two equations and solving for V_{out} gives

$$V_{\text{out}} = -\frac{A_0 V_{\text{in}} R_f}{A_0 R_i + R_f}$$

or

$$A = -\frac{A_0 R_f}{A_0 R_i + R_f} \tag{9.8}$$

One should verify that equation 9.8 reduces to equation 9.6 for $A_0 \gg R_f/R_i$ and that the amplification is given by $A = A_0$ for $A_0 \ll R_f/R_i$.

It should be noted that the results of equations 9.7 and 9.8 are still approximations, since the input resistance r_{in} and the output resistance r_{out} have been ignored in the ideal op amp representation. It turns out that the approximations are extremely good, however, so long as the external resistors are not chosen too casually. As a rule of thumb, the feedback resistor (R_2 or R_f) should be $\gtrsim r_{out}$ but $\lesssim r_{in}$. Values the order of 100 Ω to 100 kΩ are typical.

The usefulness of negative feedback cannot be overemphasized. Since all active devices are inherently nonlinear, it is essential to be able to construct circuits in which the amplification is determined by the ratio of two resistors rather than by the characteristics of the device itself. Resistors tend to be extremely linear and stable compared with nearly all other electronic components. The strategy with op amps is to provide the user with a device capable of much larger amplification than can reasonably be used, so that most of the available amplification can be traded for improved linearity. In this way circuits with extraordinarily good linearity can be constructed.

Note that we have now come full circle. The book began with linear circuits. But to make amplifiers and other useful circuits generally requires active devices that are usually quite nonlinear. Negative feedback provides the means for constructing such circuits while preserving the desired linearity.

Negative feedback also provides the circuit designer with a powerful tool for adjusting the input and output resistance of an amplifier circuit. For example, in the common emitter amplifier described in section 8.4, the emitter resistor not only reduces the amplification but also increases the input resistance of the amplifier to a value much higher than it would have otherwise been.

Consider the voltage feedback case of figure 9.4(a). If the op amp is ideal, the input resistance would be infinite. With a real op amp one is tempted to conclude that the input resistance R_{in} of the circuit would be equal to the input resistance r_{in} of the op amp itself. That this is not the case can be seen by examining the linear equivalent circuit of figure 9.6(a), in which the output resistance r_{out} is neglected. The input current I_i can be calculated as follows:

$$I_i = \frac{V_{in} - V_-}{r_{in}} = \frac{V_{out}}{A_0 r_{in}} = \frac{A V_{in}}{A_0 r_{in}}$$

The input resistance is thus given by

$$R_{in} = \frac{V_{in}}{I_i} = \frac{A_0 r_{in}}{A} \tag{9.9}$$

This equation illustrates the way in which the large available amplification serves to increase the input resistance even beyond the already large value of r_{in}.

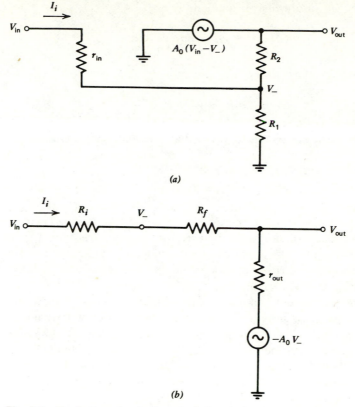

(a)

(b)

Fig. 9.6 Equivalent circuits for calculating the input resistance of the circuits in figure 9.4. (a) Voltage feedback. (b) Operational feedback.

For the operational feedback case of figure 9.4(b), the existence of the virtual ground at the inverting input makes the calculation of the input resistance especially simple. The input current I_i is just V_{in}/R_i, and the input resistance is thus

$$R_{in} = R_i$$

It appears that the input resistance can be made as large or as small as desired without limit. It is true that it can be made arbitrarily large (although it may be difficult simultaneously to achieve a high amplification), but there is a lower limit imposed by the finite output resistance of the device. The input resistance for such a case can be calculated, using the equivalent circuit of figure 9.6(b) in which the input resistance r_{in} has been neglected. The input current is

$$I_i = \frac{V_- - V_{out}}{R_f}$$

The output voltage is

$$V_{out} = I_i r_{out} - A_0 V_-$$

Eliminating V_{out} in the above equations gives

$$I_i = \frac{(A_0 + 1)V_-}{R_f + r_{out}}$$

The input resistance is thus

$$R_{in} = R_i + \frac{V_-}{I_i} = R_i + \frac{R_f + r_{out}}{A_0 + 1}$$

Since A_0 is nearly always much greater than one, the input resistance can be written as

$$R_{in} \simeq R_i + \frac{R_f + r_{out}}{A_0} \qquad (9.10)$$

For the lowest possible input resistance, one would take $R_i = 0$. Then, even with R_f considerably larger than r_{out}, an input resistance much less than $1\ \Omega$ can easily be achieved because of the A_0 in the denominator of equation 9.10. Thus negative feedback can be used either to raise the input resistance to a very high value or to reduce it to a very low value.

The output resistance of a circuit with an ideal op amp is zero. With a real op amp the output resistance R_{out} depends on the internal output resistance of the device r_{out}, but negative feedback can be used to reduce R_{out} to a very low value. Consider the equivalent circuit for the voltage feedback case in figure 9.4(a) shown in figure 9.7(a) in which the output has been shorted to ground. With the output short-circuited, the voltage V_- is zero, and the short-circuit output current is

$$I_{sc} = \frac{A_0 V_{in}}{r_{out}} = \frac{A_0 V_{out}}{A r_{out}}$$

where V_{out} is the open circuit output voltage. The output resistance is thus

$$R_{out} = \frac{V_{out}}{I_{sc}} = \frac{A r_{out}}{A_0} \qquad (9.11)$$

The large available amplification can be used to reduce the output resistance, just as it increased the input resistance (equation 9.9).

In a similar fashion, the short-circuit current for the operational feedback case of figure 9.4(b) can be calculated using the equivalent circuit in figure 9.7(b). The short-circuit output current is given by the superposition theorem as

$$I_{sc} = \frac{V_-}{R_f} - \frac{A_0 V_-}{r_{out}}$$

From the voltage-divider relation,

$$V_- = \frac{R_f V_{in}}{R_i + R_f} = \frac{V_{in}}{|A| + 1}$$

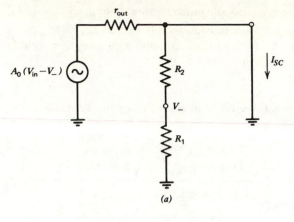

(a)

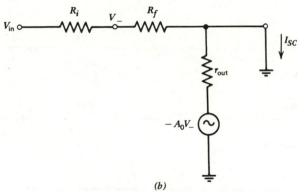

(b)

Fig. 9.7 Equivalent circuits for calculating the output resistance of the circuits in figure 9.4. (*a*) Voltage feedback. (*b*) Operational feedback.

Combining the above two equations and solving for I_{sc} gives

$$I_{sc} = \left(\frac{1}{R_f} - \frac{A_0}{r_{out}} \right) \frac{V_{in}}{1 + |A|}$$

$$= - \left(\frac{r_{out} - A_0 R_f}{R_f r_{out}} \right) \frac{V_{out}}{|A| + A^2}$$

where V_{out} is the open-circuit output voltage. The output resistance is thus

$$R_{out} = \frac{V_{out}}{I_{sc}} = - \frac{R_f V_{out}}{r_{out} - A_0 R_f} \, (|A| + A^2)$$

For the usual case in which $r_{out} \ll A_0 R_f$,

$$R_{out} \simeq \frac{|A| + A^2}{A_0} \, r_{out} \tag{9.12}$$

The output resistance R_{out} is usually much smaller than r_{out} (if $|A| \ll \sqrt{A_0}$), but at large amplification ($|A| \gg \sqrt{A_0}$) it can considerably exceed r_{out}.

The properties of the amplifier circuits with negative feedback shown in figure 9.4 are summarized in table 9.1.

TABLE 9.1 Properties of Amplifier Circuits with Negative Feedback

	Voltage Feedback Figure 9.4(a)	Operational Feedback Figure 9.4(b)
Voltage amplification (A)	$\dfrac{A_0(R_1 + R_2)}{A_0 R_1 + R_2}$	$-\dfrac{A_0 R_f}{A_0 R_i + R_f}$
($A_0 \to \infty$)	$1 + \dfrac{R_2}{R_1}$	$-\dfrac{R_f}{R_i}$
Input resistance (R_{in})	$\dfrac{A_0}{A} r_{in}$	$R_i + \dfrac{R_f + r_{out}}{A_0}$
($A_0 \to \infty$)	∞	R_i
Output resistance (R_{out})	$\dfrac{A}{A_0} r_{out}$	$\dfrac{|A| + A^2}{A_0} r_{out}$
($A_0 \to \infty$)	0	0

Two applications of op amps which are special cases of the circuits in figure 9.4 are shown in figure 9.8. The circuit in figure 9.8(a) is called a **voltage follower**. It is a case of voltage feedback with $R_1 = \infty$ and $R_2 = 0$. Like the cathode follower, source follower, and emitter follower, it has the property of near unity voltage gain ($A = 1$), high-input resistance ($R_{in} = A_0 r_{in}$) and low-output resistance ($R_{out} = r_{out}/A_0$).

The circuit in figure 9.8(b) is called a **current-to-voltage converter**. It is a case of operational feedback with $R_i = 0$. The output voltage is given by

$$V_{out} = - I_{in} R \tag{9.13}$$

One may wonder why an op amp is required at all, since a resistor by itself is also a current-to-voltage converter. The point is that the input resistance of the op amp circuit is very low ($R_{in} \simeq R/A_0$) so that, just like an ideal ammeter, it can be inserted in series with the branch whose current is to be converted to a voltage without perturbing the circuit. Or, stated another way, the voltage that can be obtained at the output for the same perturbation to the circuit is a factor of A_0 larger when an op amp is used instead of a resistor by itself.

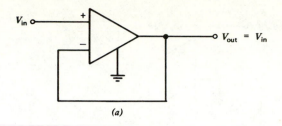

(a)

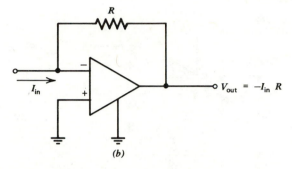

(b)

Fig. 9.8 Special cases of feedback amplifiers. (*a*) Voltage follower. (*b*) Current-to-voltage converter.

9.3 Operational Amplifier Applications

Although the operational amplifier can be used as a substitute for a single transistor or vacuum tube in any of the circuits described in the previous two chapters, it has a much wider range of application. In this section we will discuss how the op amp can be used to perform the linear mathematical operations of addition, subtraction, integration, and differentiation. In the following discussion the op amps are assumed ideal, with $A_0 = \infty$ except where indicated otherwise.

Figure 9.9(*a*) shows the basic addition circuit. Applying Kirchhoff's current law to the virtual ground at the inverting input terminal gives:

$$\frac{V_1}{R_1} + \frac{V_2}{R_2} + \frac{V_{out}}{R_f} = 0$$

Solving for V_{out} gives

$$V_{out} = -\left(\frac{R_f}{R_1} V_1 + \frac{R_f}{R_2} V_2\right) \tag{9.14}$$

For the special case of $R_1 = R_2 = R_f$,

$$V_{out} = -(V_1 + V_2) \tag{9.15}$$

In a similar manner, such a circuit can be used to add three or more voltages. The inverting input is a virtual ground and is called the **summing point**. Currents can

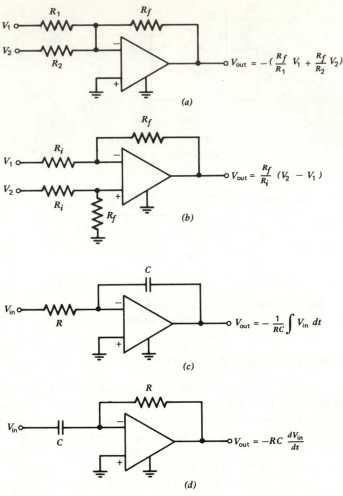

Fig. 9.9 Basic op amp circuits. (*a*) Adder. (*b*) Subtractor. (*c*) Integrator. (*d*) Differentiator.

be added by simply omitting the resistors (except R_f) connected to the summing point.

The existence of a summing point in such a circuit is the key to its operation. Since the summing point is a virtual ground, an arbitrary number of inputs can be connected to it, and each input is independent of the others. By contrast, if the summing point were not a virtual ground but, say, a resistor to ground, then its voltage would vary in response to each of the input currents, and the other input currents would be correspondingly affected. An op-amp adder thus provides isolation between circuits whose outputs are to be added, so that each circuit is oblivious to the existence of the others.

Figure 9.9(b) shows the basic subtraction circuit. The positive input is a voltage divider, so that

$$V_+ = \frac{R_f V_2}{R_i + R_f}$$

Applying Kirchhoff's current law to the inverting input gives

$$\frac{V_1 - V_-}{R_i} = \frac{V_- - V_{out}}{R_f}$$

Using the fact that $V_- = V_+$, and solving for V_{out} gives

$$V_{out} = \frac{R_f}{R_i}(V_2 - V_1) \tag{9.16}$$

For the special case of $R_i = R_f$,

$$V_{out} = V_2 - V_1 \tag{9.17}$$

Note that the addition and subtraction circuits can be combined so as to perform operations such as $V_1 + V_2 - V_3$ with a single op amp, but the design of such circuits is more difficult, because the inverting input is not a virtual ground (see problem 9.5). Since the isolation properties of the simple adder and simple subtractor are sacrificed in such a circuit, it is customary to use two op amps when both addition and subtraction are required.

It is also useful to note that either the adder or subtractor can also serve to multiply or divide any of the inputs by a constant with an appropriate choice of the resistors. Such a circuit is nothing more than an amplifier. Multiplication or division by a constant is a linear operation. To multiply or divide one variable voltage by another is a nonlinear operation, however, and requires more advanced techniques, as described in section 9.5.

Figure 9.9(c) shows the basic integrator circuit. Applying Kirchhoff's current law to the inverting input gives

$$\frac{V_{in}}{R} + C\frac{dV_{out}}{dt} = 0$$

Solving for V_{out} gives

$$V_{out} = -\frac{1}{RC}\int V_{in}\, dt \tag{9.18}$$

This result is reminiscent of the RC integrator described in section 4.6, except that in the present case there is no requirement that RC be large, or, equivalently, that V_{out} be much less than V_{in}. Actually, with a real op amp, one does, in fact, require that V_{out} be much less than $A_0 V_{in}$. A problem with a real op-amp integrator is that it has a very large voltage gain (A_0) at low frequencies. Therefore, a small dc component of voltage at the input can drive the output to saturation. This problem can be cured either by providing a dc offset adjustment or by limiting the low-frequency gain by placing a resistor in parallel with the feedback capacitor.

Figure 9.9(*d*) shows the basic differentiator circuit. Applying Kirchhoff's current law to the inverting input gives

$$V_{out} = -RC \frac{dV_{in}}{dt} \qquad (9.19)$$

This result is reminiscent of the *RC* differentiator described in section 4.6, except that in the present case, there is no requirement that *RC* be small, or equivalently, that V_{out} be much less than V_{in}. As with the integrator, one requires only that V_{out} be much less than $A_0 V_{in}$. The op-amp differentiator also has a difficulty, in that it has a very large voltage gain (A_0) at high frequencies. The result is that a great deal of noise (see section 9.7) appears at the output, unless one reduces the high-frequency gain by placing a resistor in series with the input capacitor. Op-amp integrators and differentiators can also be made using RL circuits, but this is rarely done, because inductors tend to be larger, more expensive, and less nearly ideal than capacitors.

9.4 Analog Computers

The circuits described in the previous section can be used in various combinations to solve linear differential equations. Such an application is an example of an **analog computer**. The unknown is represented as a voltage at a point in the circuit, and its value as a function of time can be determined with an oscilloscope or similar device.

As an example, consider the following differential equation written in standard form:

$$\frac{d^2x}{dt^2} + 10 \frac{dx}{dt} - \frac{1}{3} x = 6 \sin \omega t$$

One would like to construct a circuit in which $x(t)$ appears as a voltage to be measured somewhere in the circuit. This is done by first rewriting the equation with the highest derivative on the left by itself and all other terms on the right:

$$\frac{d^2x}{dt^2} = -10 \frac{dx}{dt} + \frac{1}{3} x + 6 \sin \omega t$$

One then starts with d^2x/dt^2 as an input and generates the other lower-order derivatives by successive integration, as shown in figure 9.10. The resulting terms are then added with appropriate multiplicative constants and inversions (multiplication by -1) until the quantity on the right-hand side of the equation is generated, whereupon it is fed back to the input. For a nonhomogeneous equation such as the above, a time-dependent voltage source is required (such as the sin ωt in figure 9.10). The initial conditions for the transient solution can also be simulated by placing appropriate initial voltages on the two capacitors. In the example above, the integrators are made with a time constant of $RC = 1$ s, but this is not a necessity.

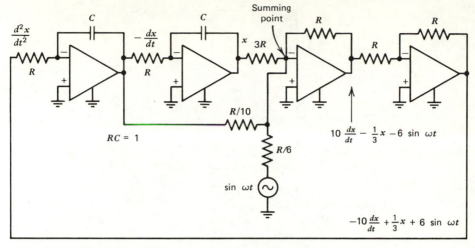

Fig. 9.10 Analog computer circuit for solving the equation
$$(d^2x/dt^2) + 10 \ (dx/dt) - \tfrac{1}{3} x = 6 \ \sin \omega t.$$

Other choices permit one to slow down or speed up the phenomenon in order to observe it on a more convenient time scale. In designing such circuits one should always check to be sure that the inverting input and the output of each op amp are connected either to resistors or capacitors, and never directly to one another or directly to a voltage source or to ground. Note that the technique described above is limited to linear equations. If the equation contained a term such as $x \, (dx/dt)$ it would be nonlinear, and no combination of the op-amp circuits discussed so far would suffice to generate a solution.

9.5 Nonlinear Operations

In addition to the linear operations described in the preceding sections, op amps can be used to perform a wide variety of nonlinear operations. In this section several such applications will be mentioned.

If a real *pn* junction diode with a *V-I* characteristic as given by equation 6.3 is used as the feedback element of an op amp, as shown in figure 9.11(*a*), the result is a device called a **logarithmic amplifier**. Applying Kirchhoff's current law to the inverting input gives

$$\frac{V_{\text{in}}}{R} = I_0 \, (e^{-eV_{\text{out}}/kT} - 1) \simeq I_0 e^{-eV_{\text{out}}/kT}$$

where the latter approximation is valid for V_{out} negative and

$$|V_{\text{out}}| \gg \frac{kT}{e} \simeq 0.026 \text{ V} \qquad \text{(at room temperature)}$$

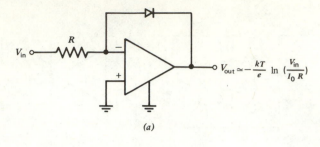

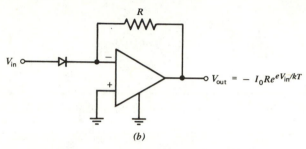

Fig. 9.11 Nonlinear op amp circuits. (a) Logarithmic amplifier. (b) Exponential amplifier.

Solving for V_{out} gives

$$V_{out} \simeq -\frac{kT}{e} \ln\left(\frac{V_{in}}{I_0 R}\right) \tag{9.20}$$

Such a logarithmic amplifier enables one to measure a voltage that varies over several orders of magnitude without having to change the range of the meter which is being used. Note that such a circuit only works for a positive V_{in}, since the log of a negative number is not real. Furthermore, the output of the circuit in figure 9.11(a) is always negative.

If the diode is used as the input element of an op amp, as shown in figure 9.11(b), the result is a device called an **exponential** (*or* **antilogarithmic**) **amplifier**. Applying Kirchhoff's current law to the inverting input terminal gives

$$\frac{V_{out}}{R} = -I_0 \left(e^{eV_{in}/kT} - 1\right) \simeq -I_0 e^{eV_{in}/kT}$$

where the latter approximation is valid for

$$V_{in} \gg \frac{kT}{e} \simeq 0.026 \text{ V} \qquad \text{(at room temperature)}$$

Solving for V_{out} gives

$$V_{out} = -I_0 R e^{eV_{in}/kT} \tag{9.21}$$

Such an exponential amplifier is easily saturated if the input voltage becomes too large. Note that such a circuit only works for a positive V_{in}, and that the output is always negative.

With the two circuits in figure 9.11, along with the linear circuits previously described, one can design circuits to perform a variety of nonlinear operations including multiplication, division, and raising a number to an arbitrary power (either positive or negative). One need only take the log of the numbers, add or subtract the logarithms, and take the exponential of the result. An example of such a circuit which produces an output proportional to the product $V_1 V_2$ (for V_1 and $V_2 > 0$) is shown in figure 9.12. One should work through the circuit stage by stage to verify that it has the predicted behavior.

The **analog multiplier** circuit described above is an example of a **single-quadrant multiplier** since, of the four possibilities, the input voltages must both be positive. With more complicated arrangements of the same basic circuits, a **four-quadrant multiplier** can be constructed in which either input voltage can have either sign. Such analog multiplier circuits are available at low cost as a single integrated circuit, as are a wide variety of circuits that perform other nonlinear operations such as division and square roots. Such circuits can be used in analog computers in the manner described in the previous section to solve nonlinear differential equations.

An entirely different kind of nonlinear operation is exhibited in the **comparator** circuit in figure 9.13(a). For such a circuit, the output is driven to saturation whenever V_1 and V_2 are different:

$$V_{out} = \begin{cases} -V_{SAT} & \text{for } V_1 < V_2 \\ +V_{SAT} & \text{for } V_1 > V_2 \end{cases} \tag{9.22}$$

For a real op amp with finite gain, the voltage difference $|V_1 - V_2|$ must exceed V_{SAT}/A_0 to saturate the output. In addition to determining the sign of a voltage, such a circuit can be used for generating square waves from a sinusoidal input, since the output switches abruptly between the two saturated levels every time the input crosses zero. Such a circuit is also called a **zero-crossing detector**.

A related circuit is the **latch circuit** shown in figure 9.13(b), which has positive feedback. With the input open circuited, it is stable only for $V_{out} = \pm V_{SAT}$. If V_{in} is momentarily made positive, V_{out} goes to $+V_{SAT}$ and remains there until V_{in} is made negative, even if the input voltage source is disconnected. Similarly, a negative V_{in} will cause the output to latch at $-V_{SAT}$. This is an example of a **bistable flip-flop** which has application as a binary memory element in digital circuits.

As a final example of the use of operational amplifiers, consider the voltage regulator circuit in figure 9.14(a). It is identical to the transistor voltage regulator shown in figure 8.9 except for the addition of one resistor (R_2) and an op amp. The Zener diode holds the noninverting input of the op amp at a constant voltage V_B. The inverting input samples a fraction f of the voltage V_L across the load resistor by means of the voltage divider potentiometer R_2. The op amp biases the base of the transistor

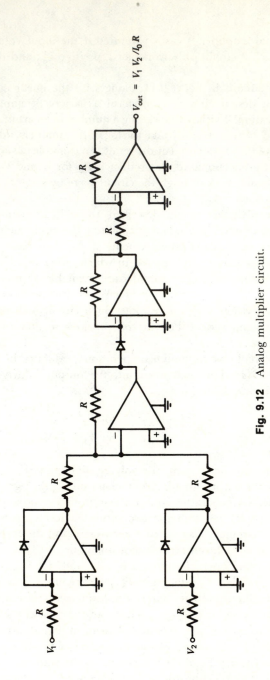

Fig. 9.12 Analog multiplier circuit.

214

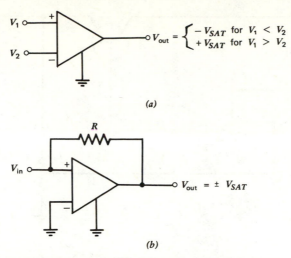

$$V_{out} = \begin{cases} -V_{SAT} & \text{for } V_1 < V_2 \\ +V_{SAT} & \text{for } V_1 > V_2 \end{cases}$$

(a)

$$V_{out} = \pm\, V_{SAT}$$

(b)

Fig. 9.13 (a) Comparator circuit. (b) Latch circuit.

in such a way that the op amp input terminals remain at the same voltage, or

$$V_L = \frac{V_B}{f}$$

By adjusting R_2 so that f ranges from zero to one, the output voltage can be regulated to any value greater than V_B.

One difficulty with this type of voltage regulator is that it works so well that it is easily damaged if connected to a load that draws excessive current. In trying to keep the output voltage constant, it will often supply enough current to destroy the load resistor, the regulator transistor, or other components in the power supply. Consequently, most general-purpose regulated power supplies are provided with some form of current limiting, so that the voltage is constant up to some maximum output current and then decreases as required to maintain a constant current. Usually the current limit is adjustable. By setting the voltage high and the current limit low, such a power supply can be made to behave much like an ideal current source. A power supply with this provision is said to be **short-circuit protected**.

Figure 9.14(b) shows how the circuit in figure 9.14(a) can be modified to include current limiting. Its operation is identical to that in figure 9.14(a) as long as the voltage drop across R_4 is sufficiently small that transistor T_2 does not conduct appreciably ($V_{BE} < 0.6$ V for silicon). If the current in R_4 rises too much, T_2 begins to conduct and reduces the base current and hence emitter current in T_1. The value of R_4 thus controls the maximum output current. Voltage regulators, incorporating most of the components shown in figure 9.14(b), are available at low cost as a single integrated circuit.

Note that a regulated power supply with current limiting is not a linear device over its entire range of operation. At low-output currents, its internal resistance is

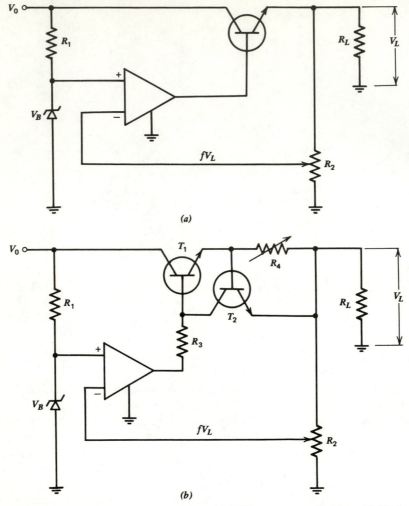

Fig. 9.14 Op amp voltage regulators. (*a*) Without current limiting. (*b*) With current limiting.

very low (dV/dI is small). At high-output currents, its internal resistance is very large (dV/dI is large). The internal resistance changes abruptly at the point at which the power supply delivers the maximum power to the load.

9.6 Amplifier Limitations

There are many other ways in which op amps, and, indeed, any amplifier fall short of ideal behavior. For example, in addition to the finite input resistance previously discussed, any amplifier will have a certain input capacitance. A typical op amp might have an input capacitance in the right range of a few to about a hundred picofarads. Since the input of an amplifier is not purely resistive, one often speaks of the **input impedance**, which, of course, is a function of frequency and has both resistive and reactive components. Similarly, the **output impedance**, especially at high frequencies, may not be purely resistive or independent of frequency.

The input impedances previously discussed are impedances between the two input terminals of the amplifier. There is also an additional impedance between each input and ground. This is called the **common mode input impedance**, and its value is typically 10–1000 times larger than the differential input impedance for most op amps. As with a simple difference amplifier (see section 8.8), an op amp will have a finite common-mode-rejection ratio, usually in the range $\sim 10^4$–10^5 (80–100 dB). This means that if a 1-V signal is applied to both inputs of the op amp simultaneously and the op amp has an open-loop voltage gain of 10^4, an unwanted output signal of ~ 0.1–1 V will result.

For an ideal op amp, the inverting input is normally operated as a virtual ground, and the voltage difference between the input terminals is negligibly small. For a real op amp, even in the absence of an input signal, the voltage difference between the two inputs may amount to a few mV. This is referred to as the input **offset voltage**. Because of the large amplification of an op amp, even a small input offset voltage can cause an objectionably large dc component at the output. Many op amps provide an extra pair of terminals to which one can apply voltages in the proper ratio in order to adjust for zero output in the absence of an input signal. Alternately, an external circuit can be added so that, for example, the voltage at the normally grounded input can be adjusted to compensate for the input offset.

In a similar fashion, with the inputs shorted together, an input **offset current**, typically in the nonampere range will produce an output voltage. The offset current tends to be less of a problem and can be easily corrected by adding an appropriate current of the opposite polarity at the summing point. The offset current, unlike the offset voltage, is temperature sensitive, and so frequent readjustment may be required in those special cases where it is large enough to be objectionable.

For any amplifier, there is a frequency above which the amplification falls significantly below its value at low frequencies. This decrease is caused largely by stray capacitance. Figure 9.15 shows a plot of the open-loop voltage gain A_0 versus frequency for a typical op amp. Such a graph is called a **Bode plot**. For many cases the gain falls by 20 dB per decade at high frequencies, until a frequency is reached at which $A_0 = 1$. The figure of 20 dB per decade is just what one would expect for a simple RC low-pass filter (see section 4.5). Similarly, the phase shift between the output and the input rises from 180° at low frequencies to 270° at high frequencies for such a case. The frequency at which A_0 falls to $1/\sqrt{2}$ of its value at zero frequency is called the **open-loop bandwidth**, Δf, of the amplifier. Note that for the case shown in figure 9.15, the open-loop bandwidth is quite small (~ 10 Hz). The frequency at which A_0 falls to one is called the **unity gain crossover frequency**. This frequency is typically in the megahertz range for most op amps.

With negative feedback, the amplification is reduced and the bandwidth is in-

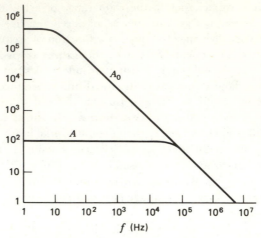

Fig. 9.15 Voltage gain of a typical op amp without feedback (A_0) and with negative feedback (A).

creased as shown by curve A in figure 9.15. It is usually possible to trade off gain and bandwidth in an amplifier circuit. For a case as in figure 9.15 in which the open-loop gain falls by 20 dB per decade, the gain-bandwidth product is constant:

$$A\Delta f = \text{constant} \tag{9.23}$$

For an amplifier with a very narrow bandwidth, such as might be used to amplify sine waves of a constant frequency, it is easy to get a very large amplification. Note that for the 20 dB/decade case shown in figure 9.15, the gain-bandwidth product is numerically equal to the unity gain crossover frequency.

Various techniques can be used to increase the gain-bandwidth product of an amplifier. A common example is the use of a compensating capacitor to provide some positive feedback which increases with increasing frequencies. The result is usually to produce a Bode plot in which A_0 remains high to a larger frequency but then falls at a rate in excess of 20 dB/decade. Up to a point such techniques can be useful, but a practical difficulty often arises. The sharp fall in A_0 versus frequency is inevitably accompanied by a large phase shift in the op amp, such that the net feedback becomes positive at some frequency, resulting in instability or oscillation (see Chapter 10). As a general rule, stable operation will result if the slope of the Bode plot is less than ~ 30–40 dB/decade at the point at which the curve without feedback merges with the curve with feedback (see figure 9.15).

In addition to its frequency-response limitation, an op amp is limited to a certain **slew rate**. This is a measure of how fast the output voltage can change, and it is a function only of the internal circuitry of the op amp. A typical slew rate is $\sim 1\text{V}/\mu\text{s}$. If the combination of input signal amplitude and frequency is such as to try to drive the output beyond this value, the output will become distorted. Sine or square waves

218 Operational Amplifiers

applied at the input will appear as triangular waves at the output with a slope equal to the slew rate.

Also, the maximum current that can be drawn from the output of an op amp is usually limited to a value somewhat less than the maximum output voltage divided by the internal output resistance. Values of 10–100 mA are typical. Most op amps are designed to be protected from short circuits, so that no damage is done if their output is inadvertently connected to ground.

9.7 Noise

Even in the absence of any input, an amplifier will produce a certain amount of noise at its output. This noise contains a broad spectrum of frequencies and is noticeable as hiss at the output of an audio amplifier. One fundamental cause of noise is the thermal fluctuation of the electrons in a resistor, which gives rise to a voltage at the resistor terminals. The magnitude of this voltage can be estimated using the equipartition theorem of statistical mechanics, which states that there is $\frac{1}{2}kT$ of energy per degree of freedom for a physical system in thermal equilibrium at temperature T. The stray capacitance associated with any real resistor, thus, on the average, stores an amount of energy given by

$$\tfrac{1}{2}C\overline{V^2} = \tfrac{1}{2}kT$$

Since the bandwidth Δf of the parallel RC circuit is given by

$$\Delta f = \frac{1}{2\pi RC}$$

the mean square noise voltage can be written in a form that is independent of C:

$$\overline{V^2} = 2\pi kTR\Delta f$$

A more exact calculation gives the result

$$V_{\text{rms}} = \sqrt{4kTR\Delta f} \tag{9.24}$$

where $kT = 4.14 \times 10^{-21}$ J at room temperature (~ 300 K). When written in this form, the noise voltage depends on the bandwidth of the instrument (oscilloscope, etc.) which is used to make the measurement. This noise is variously called **thermal noise**, **Johnson noise**, or **Nyquist noise**. It is an example of **white noise**, since all frequency components are present, just as in the case of white light. It is important to note that the noise voltage obeys a Gaussian probability distribution:

$$P(V) \sim e^{-\frac{1}{2}(V/V_{\text{rms}})^2}$$

so that voltage spikes of several times the rms value will occasionally occur. An amplifier with a bandwidth of 1 MHz and 1 MΩ input resistance will have an rms thermal noise voltage of $\sim 1.3 \times 10^{-4}$ V at its input. If it has an amplification of 1000, the noise voltage at the output would be ~ 0.13 V. Other sources of noise are usually also present, which preclude an approach to the Johnson noise limit. In any case, it is desirable to design an amplifier with the minimum permissible bandwidth in order to achieve the maximum possible signal-to-noise ratio. Sometimes amplifiers are cooled to temperatures near absolute zero ($-273°C$) in order to improve the signal-to-noise ratio.

A second form of noise, called **shot noise**, arises from the fact that electrical currents consists of the cumulative motion of many individual electrons. In a time Δt, during which n electrons cross a surface, there will be an rms fluctuation in n given by by

$$\Delta n = \sqrt{2n}$$

The corresponding rms fluctuating current is thus

$$i_{rms} = \frac{e\Delta n}{\Delta t} = \frac{e\sqrt{2n}}{\Delta t}$$

Thus, in terms of the dc current,

$$I = \frac{en}{\Delta t}$$

the noise current is

$$i_{rms} = \sqrt{2eI/\Delta t}$$

Equating Δt to the inverse bandwidth $1/\Delta f$ gives the usual expression for the rms shot noise current:

$$i_{rms} = \sqrt{2eI\Delta f} \tag{9.25}$$

The amount of shot noise usually deviates somewhat from equation 9.25 and tends to be worse in vacuum tubes and transistors than in simpler devices like resistors. Like thermal noise, shot noise is white, in that it has a constant power density per unit frequency independent of frequency.

A third type of noise, less well understood, is called **flicker noise**. It is characterized by a power density inversely proportional to frequency, so that it always dominates the other types of noise at sufficiently low frequencies. For this reason it is sometimes called **1/f noise**. The frequency at which flicker noise is comparable to the other types of noise is called the **corner frequency**, and it is typicallly ~ 1 kHz.

The absolute amount of noise produced by an amplifier is less important than the signal-to-noise (S/N) ratio at its output. A relatively large amount of noise can be tolerated if the signal is also large. However, the signal-to-noise ratio is not a useful measure of the quality of the amplifier, since much of the noise at the amplifier output may have been present at the amplifier input rather than being generated within the amplifier. Consequently, the extra noise produced by the amplifier is often expressed in terms of its **noise figure** (NF) defined by

$$NF = 10 \log_{10} \frac{(S/N)_{input}}{(S/N)_{output}} \qquad (9.26)$$

where S/N is the ratio of signal to noise power. The noise figure is always greater than 0 dB and might typically be in the range of 5–10 dB.

Noise generated at the input stage of an amplifier is usually the most troublesome, because it experiences the largest amplification. For an amplifier in which all the noise is generated by the input resistance, it turns out that the noise figure is a function only of the input resistance R_{in} and of the resistance R_s of the source which is connected to the amplifier input:

$$NF = 10 \log_{10}(1 + R_s/R_{in}) \qquad (9.27)$$

Note that when the amplifier is matched to the source $(R_{in} = R_s)$, the noise figure is 3 dB, and that the noise figure becomes very bad if the amplifier input resistance is unnecessarily low. This illustrates another advantage of constructing amplifiers with a high input resistance.

Even the noise figure is a highly imperfect measure of the quality of an amplifier, since it depends on the bandwidth, the source resistance, and the temperature of the source resistance. A better measure is to imagine a resistor with a resistance equal to the input impedance of the amplifier connected at its input. If the resistor is at a temperature of 0 K, all the noise at the amplifier output would be generated within the amplifier. If the temperature of the resistor were increased until its thermal noise just caused the amplifier output noise power to double, that temperature would be the **noise temperature** of the amplifier. An advantage of noise temperature is that it is independent of the bandwidth of the amplifier, so that it can be used to compare amplifiers with different bandwidths. Furthermore, since the noise power is proportional to the noise temperature, the noise temperature of a complicated system can be determined by adding the noise temperatures of each component of the system. Such would not have been the case with noise figure or signal-to-noise ratio. The noise temperature can be lower, but is often considerably higher than the actual temperature at which the amplifier operates.

9.8 Circuit Isolation

In addition to the random noise that is always present in electrical circuits, other types of interference caused by unavoidable coupling to nearby circuits often plague even the most experienced circuit designer unless great care is given to the physical layout of the circuit. Although the problem is not unique to operational amplifier circuits, such circuits provide an opportunity to illustrate the principles involved. Because of the large amplifications often used in op-amp circuits, their interference problems tend to be especially severe. One should be especially wary of circuits that amplify or otherwise process low-level signals (in the millivolt range) when nearby circuits involve high voltages or currents.

Unwanted coupling can take place by three basic mechanisms: resistive, inductive, and capacitive. Each of these will be considered in turn. Consider first the standard op-amp circuit in figure 9.16(a). Now suppose that a circuit designer notes that point A is at ground potential and finds it convenient for the physical layout of the circuit to use that point as the ground return of another circuit in which a current I flows. This would be perfectly acceptable, except for the fact that the conductor between point A and ground is never ideal and will, in general, have a small resistance R. Because of this resistance, a voltage drop of IR appears in series with the input to the op amp. It would not be unusual to have such a circuit in which I is 1 A and R is 0.1 Ω, giving a voltage of 100 mV at the input. If V_{in} were 1 mV, the signal would be completely masked by the interference. If the amplifier has an amplification of 1000, the output would likely be driven to saturation. If the current I were a 60-Hz sine wave, a 60-Hz square wave would appear at the output, and the desired signal would be completely lost.

The cure in this instance is relatively simple. One would disconnect the extra circuit from point A and connect it instead directly to ground. In general, one should be extremely cautious to ensure that the only currents that flow in any part of a low-level signal circuit are those produced by the signal, and not by other sources.

The second way in which interference can be coupled to a circuit is inductively, as in a transformer. Figure 9.16(b) shows an example of inductive coupling. Suppose a nearby circuit produces a fluctuating magnetic field **B**, part of which links the input loop. From Faraday's law, a voltage is produced at the input of the op amp equal to the normal component of $d\mathbf{B}/dt$ integrated over the area of the loop. The input loop can be considered as a single turn secondary of a transformer in which an adjacent loop of the interfering circuit is the primary.

This type of coupling can never be completely eliminated, but it can be greatly reduced. First, one should separate the low-level signal circuits from other high-current circuits by as much distance as possible. Second, one should reduce the magnetic fields of high-current circuits to the lowest possible value by reducing the area of the loops in which high currents flow. In an extreme case, the amplifier circuit could be enclosed in a ferromagnetic shield. Finally, one should reduce the area of the input loop. This is generally done by twisting the input leads together or by using coaxial cable at the input. Twisting the leads not only reduces the area but tends to cancel the induced voltage by the periodic reversal of the direction of the loop. A coaxial cable allows almost no inductive coupling.

A subtle variation involving both inductive and resistive effects is the **ground loop** as shown in figure 9.16(c). The fact that the input circuit is grounded at two different points seems perfectly innocuous until one considers that this forms a loop, part of which is common to the input circuit. A fluctuating magnetic field **B** will produce currents in this loop which will, in turn, cause a voltage drop across the resistance R of the input leads. The solution is to ensure that the input circuit is grounded at only one point, preferably close to the input of the amplifier.

The third type of coupling arises from the stray capacitance C between the amplifier input and a nearby circuit having a large time-varying voltage as shown in

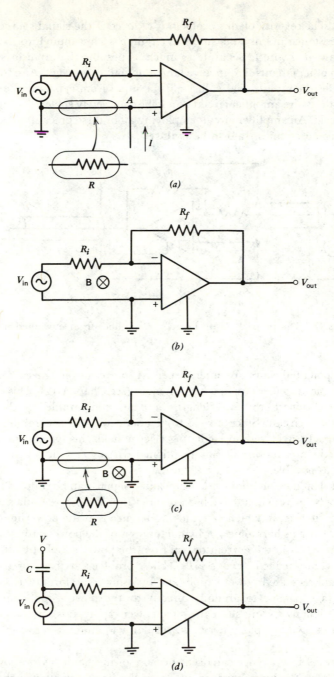

Fig. 9.16 Possible sources of interference in an amplifier circuit. (*a*) Resistive coupling. (*b*) Inductive coupling. (*c*) Ground loop (inductance + resistance). (*d*) Capacitive coupling.

figure 9.16(d). The severity of this problem is reduced if the signal source V_{in} has a low internal resistance and/or if the amplifier circuit has a low input impedance (small R_i). In any case, it is wise to separate as much as possible the input of a low-level amplifier from other circuits that involve large fluctuating voltages. Also, the use of a coaxial cable for the input signal will shield the input from capacitive coupling. In extreme cases the entire amplifier circuit should be completely enclosed in a grounded, conducting shield. An amplifier circuit employing all of these precautions is shown in figure 9.17. Such a circuit is said to be **isolated**.

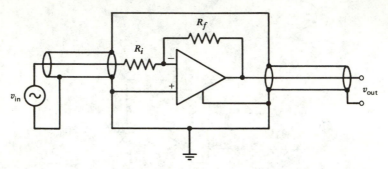

Fig. 9.17   The proper way to isolate an amplifier from unwanted interference.

A fourth potential source of interference is from electromagnetic radiation in which parts of the amplifier circuit act as antennas (see Chapter 12). This happens at high frequencies where the wavelength becomes comparable to the physical dimensions of the circuit. Since an electromagnetic wave is a combination of an oscillating magnetic field and an oscillating electric field, this type of interference is minimized by the precautions already mentioned. In any case, it is wise to keep all leads as short as possible.

Finally, it should be mentioned that unwanted signals can be coupled in through the power-supply leads, which have been ignored thus far. These signals are caused by inadequate filtering, in which case the interference is at the powerline frequency (usually 60 Hz) and its harmonics, or by fluctuations of the power-supply voltages in response to variations in the current drawn from the power supply by other circuits which may be connected to it (**cross-talk**). These problems are best cured by using a well-filtered and regulated power supply and by using some form of low-pass filter (perhaps just a capacitor to ground) directly at the power-supply input to the amplifier. When many circuits share the same power supply, it is common to provide each one with its own low-pass filter or IC regulator to minimize cross-talk through the power supply.

Other types of interference can occur because of mechanical vibrations of parts of the circuit (called **microphonics**), in which the stray capacitance changes in a time-dependent fashion. Vacuum-tube circuits are particularly susceptible to microphonics because of the small delicate grids and the relatively high voltages. Similarly, certain types of dielectrics such as the insulation often used in coaxial cables will develop a small voltage in response to applied pressure. Such materials are said to be **piezoelectric**.

These practical considerations occupy much of the attention of the professional circuit designer and are a frequent cause for the failure of an apparently well-designed circuit to perform as expected.

9.9 Summary

A dc amplifier with inverting and noninverting inputs, high input resistance, low output resistance, and large voltage gain is called an operational amplifier. Its uses are numerous. In addition to simple voltage amplifiers, op amps can be used to perform the basic linear mathematical operations of addition, subtraction, integration, and differentiation. In combination, op amps can be used as an analog computer to solve linear differential equations.

With a nonlinear component such as a *pn* junction diode, the op amp can be used to produce an output proportional either to the logarithm or exponential of the input voltage. This permits the possibility of performing nonlinear operations such as multiplying, dividing, and raising a number to an arbitrary power. Other useful nonlinear applications include the comparator and the latch, which make use of the fact that a finite voltage difference at the input will drive the output to saturation, either negative or positive. Op amps also make extremely good voltage regulators.

As useful as op amps are, they are always limited in frequency response and have a certain noise at the output. One can usually trade off gain and bandwidth, and the noise is minimized if the bandwidth is small. Op-amp circuits, along with all other electrical circuits, are susceptible to a variety of sources of interference, and considerable care in the physical layout and construction of such circuits must often be exercised.

The limitations and potential problems with the construction of op amp and other circuits are so numerous as to risk the total discouragement of the beginner in electronics. Such an attitude is not warranted, however. The problems usually come only one or two at a time, and armed with a knowledge of the common pitfalls and a proper dose of caution and persistence when things don't quite work right on the first try, even a total beginner can these days successfully build quite sophisticated electronic circuits.

Problems

9.1 Calculate the amplification of the circuit on the following page.

9.2 Consider the op-amp circuit in figure 9.4(*b*) in which $R_i = R_f$ and the op amp is real with an open-loop gain of $A_0 (\gg 1)$, an input resistance of r_{in}, and zero output resistance. How large can R_i be made if the amplification is to be given within a factor of two by equation 9.6?

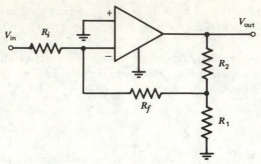

Illustration for Problem 9.1.

9.3   Calculate the input resistance of the circuit in figure 9.4(a), assuming the op amp is real, with $A_0 = 10^5$, $r_{in} = 1$ MΩ, $r_{out} = 0$, and $R_1 = R_2 = 1$ kΩ.

9.4   Calculate the output resistance of the circuit in figure 9.4(b), assuming the op amp is real, with $A_0 = 10^5$, $r_{in} = \infty$, $r_{out} = 10$ Ω, and $R_i = R_f = 10$ kΩ.

9.5   Calculate the values required for R_1, R_2, and R_3 in the circuit below in order for the output to be given by $V_{out} = -(V_1 + V_2 - V_3)$.

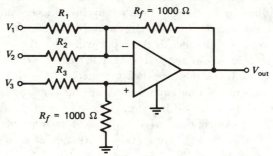

9.6   Calculate the amplification and the input resistance for the circuit below, assuming the op amp is ideal.

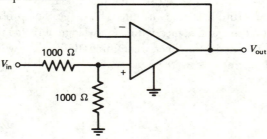

9.7   Calculate the amplification of the circuit below, assuming the op amp is ideal.

9.8   Calculate the output V_{out} in terms of the input voltage V_{in} for the following circuit in which the op amp is ideal.

9.9   Calculate the amplification as a function of frequency for the circuit below in which the op amp is ideal.

226 Operational Amplifiers

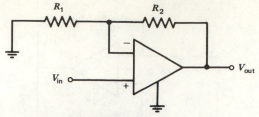

Illustration for problem 9.7.

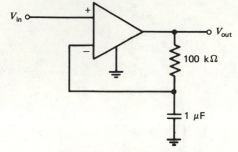

Illustration for problem 9.8.

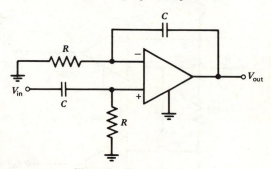

Illustration for problem 9.9.

9.10 Show that the circuit below behaves as a noninverting integrator.

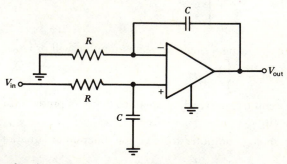

9.11 Show that the circuit below behaves like a negative resistance (i.e., I decreases as V increases), and calculate the input resistance, $R_{in} = V/I$.

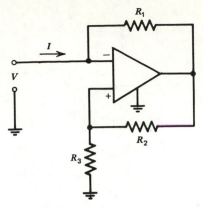

9.12   If the op amps below both saturate at an output of ± 10 V, find V_A and V_B for V_{in} between zero and $+15$ V.

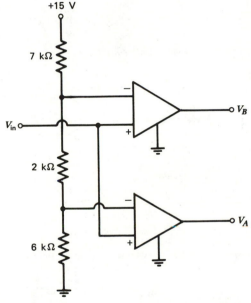

9.13   Design a circuit using a single ideal op amp that will produce an output given by $V_{\text{out}} = -7 \int V_1 \, dt - \frac{1}{4} \int V_2 \, dt$.

9.14   Design a circuit using two ideal op amps that will produce an output given by $V_{\text{out}} = 5 \int V_1 \, dt - 10 V_2$.

9.15   Design an analog computer to solve the differential equation,

$$4 \frac{d^2 x}{dt^2} - 20 \frac{dx}{dt} + 2x = 100.$$

9.16   Design a circuit using ideal op amps that will produce an output given by $V_{out} \propto \sqrt{V_{in}}$, for $V_{in} > 0$.

9.17   Show that the circuit below behaves as a logarithmic amplifier if the transistor is real. Such circuits are often used in preference to the circuit in figure 9.11 (a) because the logarithmic variation holds over a wider range of input voltages.

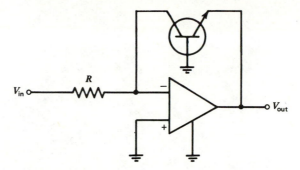

9.18   Calculate the rms Johnson noise voltage at the output of the circuit in figure 9.4(b) if $R_f = 100$ kΩ, assuming A_0 is given by figure 9.15 and $A = 100$.

9.19   Calculate the rms Johnson noise voltage at the output of the circuit below:

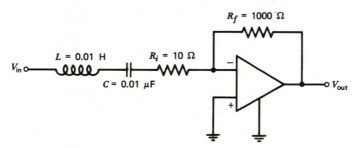

9.20   Calculate the noise figure in decibels for an amplifier with an input resistance of 1 kΩ driven by a source with a 10-kΩ output resistance.

Other Nonlinear
Circuits and Devices

10.1 Oscillators

An amplifier with positive feedback can be made to produce an output even in the absence of any input. Such circuits are called **oscillators**. They are useful for producing ac voltages of adjustable frequency from a dc source. Suppose that a fraction f of the output is returned and added to the input V_{in} of an amplifier. Then the output is given by

$$V_{out} = A_0(V_{in} + fV_{out})$$

or

$$V_{out} = \frac{A_0 V_{in}}{1 - A_0 f}$$

Even though V_{in} is zero, an output voltage can still be achieved if the condition

$$A_0 f = 1 \qquad (10.1)$$

is satisfied. Since A_0 and f usually depend on frequency, the above condition, called the **Barkhausen criterion**, is usually satisfied at only a single frequency, and that is the frequency at which the circuit will oscillate.

Figure 10.1 shows an oscillator circuit that uses an operational amplifier. The LC circuit can be considered as a resonant filter that eliminates from the amplifier input any angular frequencies significantly different from $\omega_0 = 1/\sqrt{LC}$. With a sinusoidal voltage of angular frequency ω_0 at V_1, the amplifier is alternately driven to saturation in the positive and negative direction, and so it produces a square wave at V_2. This square wave has a strong fundamental Fourier component at frequency ω_0, part of which is fed back to the noninverting input through resistor R in order to keep the oscillation from damping out even in the absence of any externally applied voltage at V_1. Such a circuit thus produces both a sine wave and a square wave output.

It is reasonable to wonder how the oscillation gets started in the first place. One might suppose that it is necessary initially to apply a sinusoidal voltage at V_1. This is seldom a problem, however, since there is always some noise present at the output. This noise has a Fourier component at frequency ω_0, and because of the positive

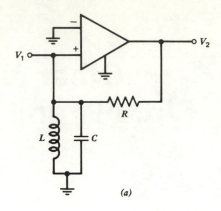

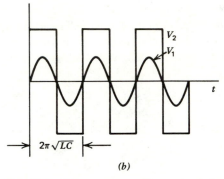

Fig. 10.1 The op amp oscillator circuit in
(*a*) produces a sinusoidal and a square wave
output as shown in (*b*).

feedback, it rapidly grows in amplitude (in just a few cycles, depending on the strength of the feedback) until the output amplitude saturates. In practice, the problem is often just the opposite. Circuits designed as amplifiers, especially if they have high gain and large bandwidth, often have enough stray capacitance to produce the positive feedback. Great care must be exercised in the construction of high-gain amplifiers to ensure that unwanted oscillations do not occur.

An oscillator circuit need not use op amps. Any device that can be used as an amplifier, such as a vacuum tube, FET, or bipolar transistor, can also be used as an oscillator. However, a single vacuum tube or transistor, when used in a circuit that provides a large voltage amplification, will generally produce an output that is 180° out of phase with the input, and so an additional phase shift of 180° is required to achieve positive feedback. For example, figure 10.2(*a*) shows an FET **Hartley oscillator**. A Hartley oscillator achieves the 180° phase shift desired for positive feedback by means of a tapped inductor in the gate circuit. The phase of the voltages at the two ends of the inductor differ by 180° with respect to the grounded tap on the

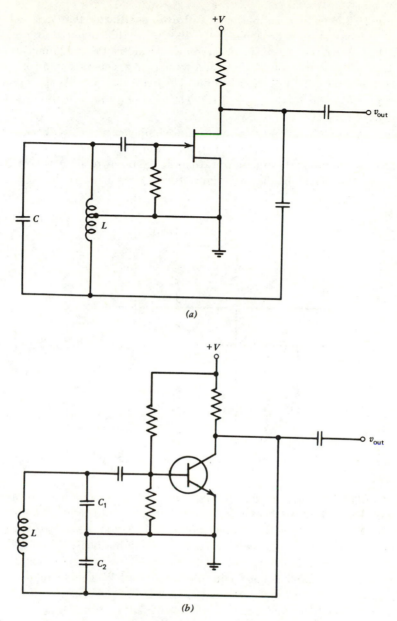

Fig. 10.2 Typical oscillator circuits. (*a*) Hartley. (*b*) Colpitts.

inductor. The situation is analogous to the full-wave rectifier with a center-tapped transformer (see figure 6.7). The frequency of the Hartley oscillator is determined almost entirely by the value of $1/\sqrt{LC}$.

Figure 10.2(*b*) is an example of a **Colpitts oscillator** that uses a bipolar transistor. A Colpitts oscillator achieves the $180°$ phase shift required for oscillation by using the fact that the two capacitors are in series, and so the circulating ac current in the LC circuit produces voltage drops across the two capacitors that are of opposite sign relative to ground at any instant of time. The frequency of the Colpitts oscillator is determined primarily by the value of $1/\sqrt{LC}$, where C is the series combination of C_1 and C_2.

Many other combinations of active circuit components and feedback methods are frequently encountered. Furthermore, it is not necessary that oscillators contain LC circuits. One particularly straightforward although rarely used type of oscillator is the RC phase-shift oscillator shown in figure 10.3. Use is made of the fact that the

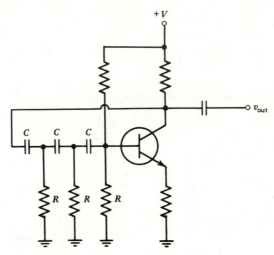

Fig. 10.3 *RC* phase shift oscillator

output of an *RC* filter differs in phase from the input by an amount that can vary from zero to $90°$ (see figure 4.9). Therefore, two *RC* filters can just produce the required $180°$ phase shift, but only in the limit of infinite attenuation. Therefore, such phase shift oscillators normally use three *RC* sections, each with a phase shift of $60°$. Even then, significant attenuation of the feedback signal occurs, requiring that the amplifier have appreciable voltage gain (see problem 10.3). The angular frequency of oscillation is on the order of $1/RC$, but calculation of the exact frequency is not trivial, because consideration must be given to the fact that the output of each *RC* circuit is loaded by the input of the next (see problem 10.2). In practice, *RC* phase-shift oscillators are only useful at audio frequencies and below ($f \lesssim 10$ kHz), because at higher frequencies stray capacitance and extraneous phase shifts become too important to neglect.

A particularly stable form of oscillator uses a piezoelectric crystal of quartz in place of the LC circuit. The symbol for a quartz crystal is shown in figure 10.4(*a*).

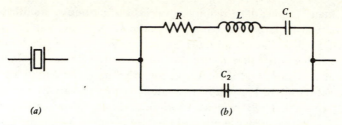

(a)   (b)

Fig. 10.4 Quartz crystal. (*a*) Symbol. (*b*) Electrical equivalent circuit.

Such a crystal exhibits a high Q resonance when a sinusoidal voltage of the appropriate frequency is applied between its faces. The frequency is determined almost entirely by the thickness of the crystal. Although the quartz crystal is an electromechanical device, its behavior can be described by an electrical equivalent circuit as shown in figure 10.4(*b*). The R, L, and C_1 represent the series mechanical resonance. The C_2 represents stray capacitance in the crystal holder and leads. The ratio of L (many henries) to C_1 ($\ll 1$ pF) is much higher than could be achieved with real inductors and capacitors. Quartz crystals are available with resonance frequencies from a few kHz to about 100 MHz.

A typical crystal oscillator circuit using an op amp is shown in figure 10.5. The

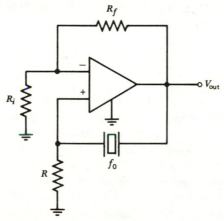

Fig. 10.5 Op-amp crystal oscillator circuit.

circuit resembles that in figure 10.1(*a*), except that the series resonance of the crystal is used instead of the parallel resonance of an LC circuit to provide positive feedback at the desired frequency. Crystal oscillators that maintain a constant frequency to better than one part in 10^6 are not at all uncommon. As a result, quartz crystals are useful as time and frequency standards, and are even found in some types of wristwatches.

Whereas crystal oscillators are excellent in applications where a fixed frequency is required, many applications require a **variable-frequency oscillator** (VFO). One straightforward approach is to make the inductor or, more usually, the capacitor in an LC-controlled oscillator adjustable. Such an approach is satisfactory if sufficient care is exercised in construction to ensure adequate frequency stability in the presence of mechanical vibrations and if rapid, precise, automatic, or remote control of the frequency is not required. For relatively small frequency variations, use can be made of varicaps (see section 6.8) in place of or in addition to the capacitor in an LC oscillator so that the frequency can be adjusted by means of a variable voltage. More sophisticated integrated circuits, called **voltage controlled oscillators** (VCOs) are available which provide a linear variation of frequency with applied voltage over a factor of 10 or more and which provide a selection of output waveforms (square, triangular, etc.).

Finally, it should be noted that oscillators are not the only application of positive feedback. Recall that one of the virtues of negative feedback is to increase the bandwidth of an amplifier. Not surprisingly, then, positive feedback can be used to narrow the bandwidth of an amplifier. In applications such as radio communications, in which a high degree of frequency selectivity is desired, positive feedback just below the level required for oscillation is sometimes used to increase selectively the amplification at a particular frequency. Such circuits are called **Q-multipliers**, because they result in a bandpass characteristic much sharper than would otherwise be allowed by the Q of the associated LC circuit. Quartz crystals, because of their very high Q are also often used in **crystal filter circuits** where frequency selectivity is important.

10.2 Multivibrators

The latch circuit shown in figure 9.13(b) is one example of a class of circuits known as a **multivibrator** or **flip-flop**. As the name suggests, a flip-flop is a circuit that abruptly changes from one state to another. Although multivibrators can be made with operational amplifiers, a simpler and more usual design uses a pair of transistors.

For example, the circuit in figure 10.6(a) is a bistable multivibrator very much like the op amp latch circuit previously described. A bistable multivibrator is a circuit that will remain in either of two states indefinitely until caused to change state by an externally applied signal. To understand its operation, imagine that $V_1 = 0$. Then the transistor on the right has no base current and hence no collector current, since $I_C = \beta I_B$. Therefore, all the current that flows through R_2 goes into the base of the left-hand transistor, driving it into saturation. In the saturated condition, V_1 is zero, as assumed at the outset. However, by symmetry the circuit is equally stable with $V_2 = 0$ and the right-hand transistor saturated. The circuit can be made to switch from one state to the other by simply grounding either V_1 or V_2 as appropriate. One way to think of the bistable multivibrator or latch circuit is as an oscillator with positive feedback at zero frequency. The oscillator begins in whichever state is dictated by the initial conditions, but because the period of oscillation is infinite, it never gets to the

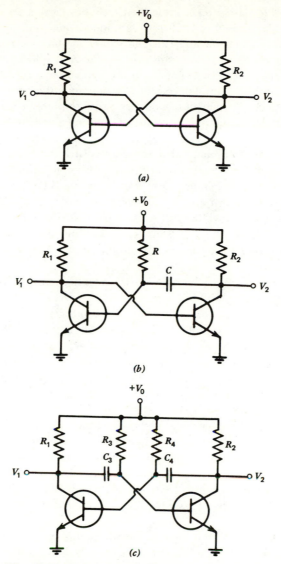

Fig. 10.6 Multivibrator circuits. (*a*) Bistable. (*b*) Monostable. (*c*) Astable.

other state. Bistable multivibrators can be used as digital memory devices (see next chapter) or as frequency dividers, since alternate pulses restore the circuit to its initial condition.

A different type of circuit, known as a **monostable multivibrator**, is shown in figure 10.6(*b*). A monostable multivibrator is a circuit that is stable in only one state. It can be put in its unstable state by an externally applied signal, but it automatically returns to its stable state after a prescribed time has lapsed. As with the previous circuit, it is stable with $V_1 = 0$. If V_2 is grounded momentarily, the capacitor C behaves transiently like a short circuit and causes the base current, and hence the collector current, of the left-hand transistor to go to zero. Then all the current in R_1 flows into the base of the right-hand transistor, holding it in saturation until the capacitor C can recharge through resistor R, whereupon the circuit switches back to its initial state. Such a circuit thus produces a square pulse of voltage at V_1 with a duration determined by the time constant RC and independent of the duration and amplitude of the pulse that caused it to change state.

Such circuits are sometimes called **one-shot multivibrators** and have a variety of uses. One use is with an integrator at the output, for generating the sweep voltage in an oscilloscope when the initiation of the sweep must be triggered by an external voltage. Another use is for producing large pulses of standard width from input pulses of varying amplitude and width. Still another use is for delaying a pulse by a known amount of time. The input pulse puts the circuit in its unstable state. By differentiating the output, a pulse can be produced at a later time when the circuit switches back to its original state.

A third type of multivibrator is the **astable multivibrator** shown in figure 10.6(c). An astable multivibrator is not stable in either state and spontaneously switches back and forth at a prescribed rate, even in the absence of any input signal. Assume that V_1 is initially at ground. The base of the right-hand transistor will also be at ground until C_3 can charge up enough through R_3 that the right-hand transistor will saturate, whereupon V_2 goes to zero, causing the base of the left-hand transistor to go to zero. Then V_1 rises to a positive value until C_4 charges up through R_4, causing the left-hand transistor to conduct, which starts the cycle all over again. The result is that the circuit automatically switches back and forth between the two states. The time spent in each state can be controlled by the time constants of the RCs in the base circuits. An astable multivibrator is basically an oscillator, but it allows some flexibility in the shape of the output waveform. The capacitors eliminate the positive feedback at dc, thereby avoiding the latch-up that occurs with the bistable multivibrator. Note the resemblance to the RC phase-shift oscillator, but remember that with multivibrators the voltages are not sinusoidal, and so the concept of phase is of limited use.

A useful variation of the monostable multivibrator is the **Schmitt trigger** circuit shown in figure 10.7(a). Since one or the other of the transistors is always in conduction, the emitter voltage V_E is approximately the same level (for $R_1 \simeq R_2$), which is a fraction of the power supply voltage V determined by the ratio of the various resistors. If V_{in} is less than V_E, the transistor on the left does not conduct and its collector rises to a high voltage, holding the right-hand transistor in conduction. Under this condition the output voltage V_{out} is equal to V_E. If V_{in} exceeds V_E, the left hand transistor begins to conduct, lowering its collector voltage and hence the base voltage of the transistor on the right. This allows V_{out} to rise to $+V$ where it remains until V_{in} drops below V_E. Actually, the input voltage must fall somewhat below V_E before the output switches (see problem 10.8). The Schmitt trigger is thus like a

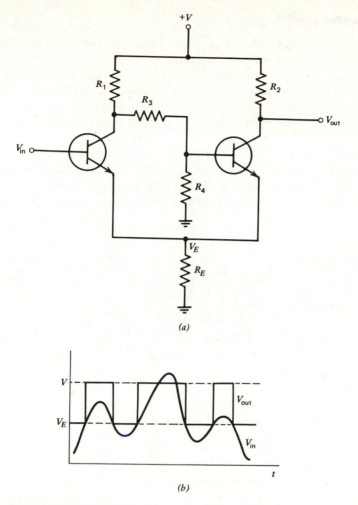

Fig. 10.7 The Schmitt trigger circuit in (a) produces an output (b) that switches between two levels whenever the input voltage crosses the lower of the two levels.

comparator (see figure 9.13), in that it switches states abruptly when the input crosses a specified value as shown in figure 10.7(b). In addition to its use as a comparator, the Schmitt trigger is useful for eliminating noise on certain types of signals and for generating square waves from a sinusoidal input. It is also used as a trigger level control in oscilloscopes to initiate the sweep when the trigger signal exceeds a certain preset level. In practice, all these circuits are usually seen with small additional capacitors whose function is to reduce the time required for the circuit to switch from one state to the other.

10.3 Tunnel Diodes

A *pn* junction diode that is heavily doped so as to increase the concentration of charge carriers has an electric field that is concentrated very near the junction. The region over which the field exists is so narrow that charges can tunnel through the barrier by a quantum mechanical effect. Such a diode is called a **tunnel diode** and has an *I-V* characteristic as shown in figure 10.8(*a*). Unlike an ordinary diode, the tunnel diode conducts strongly when reverse-biased or when forward-biased by a small amount. For large forward bias, the tunnel diode behaves like any other *pn* junction diode (i.e., $I \simeq I_0 e^{eV/kT}$).

The characteristic of the tunnel diode that sets it apart from all the other devices encountered so far is the multivaluedness of the current. As the current is increased from zero, the voltage increases until it reaches point 1 in figure 10.8(*a*). Then it

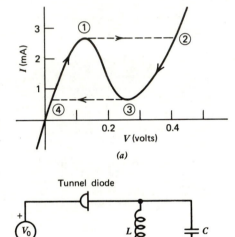

(a)

Fig. 10.8 (*a*) Typical tunnel diode characteristics. (*b*) Tunnel diode oscillator.

abruptly jumps to point 2. If the current is then decreased, the voltage falls to point 3, and then jumps to point 4. Such a device is said to exhibit **hysteresis**, since the curve does not retrace itself as the current oscillates.

A tunnel diode can be used to construct multivibrators similar to those described in the previous section. For example, if a current source of, say, 2 mA is connected to the tunnel diode whose characteristics are shown in figure 10.8(*a*), the device is stable in one of two states with different voltages. Monostable and astable multivibrators can be made in similar fashion.

The region of the curve in figure 10.8(*a*) between points 1 and 3 has $dI/dV < 0$, and hence is said to have **negative resistance**. A negative-resistance device can be used as part of an oscillator circuit, as shown in figure 10.8(b). The dc voltage V_0 establishes an operating point in the negative resistance region. The ac linear equivalent circuit then consists of a negative resistance:

$$R = \frac{dV}{dI}\bigg|_{V=V_0} \qquad (10.2)$$

in parallel with an LC. The differential equation for such a circuit predicts a sinusoidal oscillation that grows rather than damps exponentially, as was the case with the transient RLC circuits described in Chapter 3. The oscillation eventually reaches a limiting amplitude when the resistance departs significantly from the value in equation 10.2, and the diode is just able to compensate for losses in the nonideal L and C. It is a general feature of systems that are unstable in their linear (small amplitude) limit to grow exponentially in time until some nonlinear effect terminates the growth. The growth is, then, a transient state that often goes unnoticed, and the amplitude of the steady-state oscillation is determined entirely by the nonlinearities.

In addition to its simplicity, a tunnel diode also has an advantage in its fast switching speed. Charge carriers cross the junction at essentially the speed of light, in contrast to the slow diffusion of charges in the bipolar transistor. For this reason, multivibrators using tunnel diodes are ideal for high-speed digital computers, and tunnel diode oscillators have been made to operate at frequencies as high as 10^{11} Hz.

A close relative of the tunnel diode is the **back diode**. By controlling the doping during manufacture, it is possible to suppress the peak forward current [point 1 in figure 10.8(a)] while retaining the rapid rise in current in the reverse direction. For small voltages (<0.6 V for silicon), such a diode thus behaves just the opposite of an ordinary diode, except that the knee of the V-I characteristic occurs very close to zero voltage. Back diodes are therefore useful for rectifying very small ac voltages where an ordinary diode would simply behave like a high-value resistor.

Negative-resistance devices were known well before the advent of modern semiconductor technology. For example, the **neon bulb** in which two electrodes are sealed in a glass envelope filled with low-pressure neon gas is widely used in pilot lamps to indicate when a particular circuit is energized, but it is also known to exhibit negative resistance. The neon bulb, whose symbol is shown in figure 10.9(a), draws essentially no current until the voltage across its terminals increases to about 80 V, whereupon the neon becomes ionized, the voltage drops back to a lower value, and the current increases to whatever value is required to maintain the voltage at $\sim$60 V. In normal use, the neon bulb must therefore be used with a voltage greater than $\sim$80 V and a series, current-limiting resistor.

A simple circuit, called a **relaxation oscillator**, using a neon bulb, is shown in figure 10.9(b). When the circuit is turned on, the capacitor C begins to charge through resistor R, exponentially approaching the voltage V_0, as shown in figure 10.9(c). Until the capacitor voltage reaches $\sim$80 V, no current flows in the neon bulb, but when the neon bulb finally begins to conduct, it rapidly discharges the capacitor to $\sim$60 V, whereupon the neon bulb goes out (provided R is sufficiently large) until the capacitor is able to recharge to $\sim$80 V. The voltage across the neon

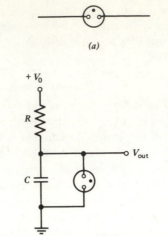

(a)

(b)

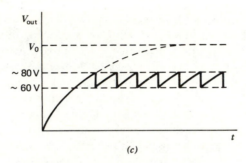

(c)

Fig. 10.9 The neon bulb (*a*), because of its negative resistance characteristic, can be used in a relaxation oscillator circuit (*b*) to produce a sawtooth output voltage (*c*).

bulb thus consists of a dc component of $\sim$70 V and a sawtooth-shaped ac component with a peak-to-peak amplitude of $\sim$20 V. Such oscillators are especially simple and inexpensive, but they are not very stable and are limited to frequencies below about 100 kHz.

10.4 Unijunction Transistors

A three-terminal device with negative resistance characteristics similar to the tunnel diode is the **unijunction transistor** (UJT). It consists of a single bar of *n*-type silicon semiconductor with a small *pn* junction near its middle which forms the emitter. At each end of the bar is a base terminal made by ohmic contacts to the respective base leads. The symbol and equivalent circuit for the UJT are shown in

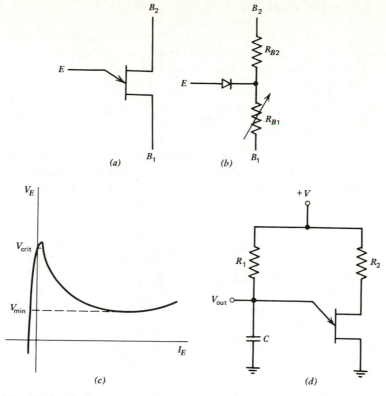

Fig. 10.10 Unijunction transistor. (*a*) Symbol. (*b*) Equivalent circuit. (*c*) Emitter characteristics. (*d*) Relaxation oscillator.

figure 10.10(*a*) and (*b*), respectively. With the emitter open-circuited, the resistance between the bases is typically a few thousand ohms, with R_{B1} somewhat greater than R_{B2}. If base 1 is grounded, a voltage V_E applied at the emitter has no effect unless it exceeds a value given by

$$V_{\text{crit}} = \frac{R_{B1}}{R_{B1} + R_{B2}} V_{B2}$$ (10.3)

whereupon the diode begins to conduct and current flows into the emitter. This current causes R_{B1} (and hence V_E) to decrease and I_E to increase. This decrease in V_E as I_E increases is the origin of the negative resistance characteristic of the UJT. The emitter characteristic of a typical UJT is shown in figure 10.10(*c*).

Like the tunnel diode, the UJT can be used as an oscillator. Figure 10.10(*d*) shows a typical UJT oscillator. As capacitor C charges through R_1, the emitter voltage will increase until the emitter begins to conduct, whereupon the capacitor will suddenly discharge, and the voltage will drop below the minimum value shown in figure 10.10(*c*). Then the emitter ceases to conduct (I_E drops to a very low value), and the cycle begins again. The output voltage across the capacitor thus has a sawtooth waveform. The operation is analogous to the neon bulb relaxation oscillator except that the waveform returns nearly to zero when the capacitor discharges. An additional output from base 2 would provide negative-going spikes every time the capacitor discharges. Such a circuit is another example of a relaxation oscillator, and it finds use in the same applications as the astable multivibrator discussed earlier.

10.5 Silicon-Controlled Rectifiers

Another three-terminal nonlinear device is constructed from a *pnpn* structure, as shown in figure 10.11(*a*). Such a device is called a **silicon-controlled rectifier** (SCR) or a **thyristor**. The symbol for an SCR is shown in figure 10.11(*b*). The SCR can be represented as two bipolar transistors connected as in figure 10.11(*c*).

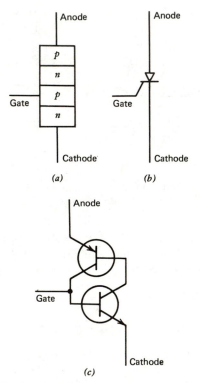

Fig. 10.11 Silicon-controlled rectifier. (*a*) Structure. (*b*) Symbol. (*c*) Representation in terms of bipolar transistors.

To understand its operation, assume first that no current flows into the gate (gate either open-circuited or connected to the cathode). Then the bottom transistor is biased to cutoff and hence draws no collector current. The upper transistor then has no base current, and it is also an open circuit. The device thus looks like an open circuit between anode and cathode for either polarity of voltage. Now suppose a positive current flows into the gate. The bottom transistor begins to conduct, causing a base current to flow into the upper transistor, causing it to conduct also. It then produces an additional base current in the lower transistor, causing it to conduct even more, and so forth, until both transistors are biased to saturation. The device then remains a short circuit between anode and cathode, even when the gate current is removed. It can be made to cease conducting only by reducing the anode current below a small value called the **holding current** (or by reversing the anode-to-cathode voltage). The SCR thus behaves like a switch that can be closed (turned on) by a momentary pulse of current at its gate. The gate input resistance is quite small ($\lesssim 100 \ \Omega$), and the size and duration of the required gate trigger pulse are also small.

A variation of the SCR is the **silicon-controlled switch** (SCS), in which a fourth terminal (called the **anode gate**) is connected to the base of the upper transistor in figure 10.11 (c). This fourth electrode allows the device to be switched off by the application of a momentary positive current.

One very common use of the SCR is for controlling the ac power delivered to a load. Such a circuit is shown in figure 10.12 (a). By varying the resistor R, the fraction of the cycle over which the device conducts can be controlled, as shown in

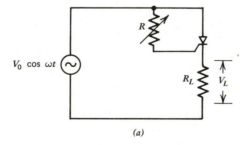

(a)

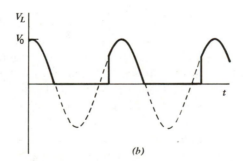

(b)

Fig. 10.12 The SCR circuit in (a) allows the average power in R_L to be varied by controlling the waveform as shown in (b).

figure 10.12(b), and hence the average power dissipated by R_L can be adjusted. The advantage of such a circuit over a simple adjustable series resistor is that essentially no power is wasted in the SCR, since it always has either a zero current or a zero voltage. The power dissipated by resistor R can be very small. The disadvantage of the SCR is that the waveform across the load R_L in no way resembles a sine wave. Often this is of no concern, and SCRs find wide application in consumer appliances such as electric-light dimmers and electric-motor speed controls.

One common application of the SCR is as a rectifier in a variable voltage power supply. Since the waveform of figure 10.12(b) resembles the output of a half-wave rectifier (see figure 6.6), the resistor R_L in figure 10.12(a) can be replaced with a low-pass filter as shown in figure 10.13, and the result will be to produce a dc output

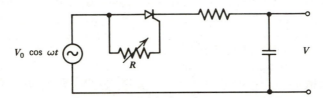

Fig. 10.13 Half-wave SCR-controlled variable power supply.

voltage whose value can be adjusted by the resistor R. Such a power supply is much more efficient than one in which control is achieved by wasting power in a series resistor or transistor. Full-wave and bridge rectifier versions of the SCR-controlled power supply can be constructed in similar fashion.

A limitation of the SCR circuits of figures 10.12 and 10.13 is that the output wave shape can be varied only between a quarter cycle and a half cycle. The average power in the load can thus be varied by only a factor or two. This limitation can be largely overcome by the addition of a capacitor, as shown in figure 10.14(a). With R small, the capacitor has very little effect, and the SCR fires shortly after the voltage crosses zero in the positive-going direction, as shown in figure 10.14(b). When R is increased, the time at which the SCR fires moves later, because a larger source voltage is required, just as in figure 10.12, but the RC circuit produces an additional phase shift that can approach 90°. Consequently, the capacitor voltage reaches its peak value just before the source voltage crosses zero in the negative-going direction, as shown in figure 10.14(c). The result is that the output waveform duration can be adjusted from nearly zero to almost half a cycle. Similarly, the voltage output of the power supply in figure 10.13 could, with such a modification, be varied from near zero to full output.

The circuit in figure 10.12 has the additional disadvantage of conducting at most over a half a cycle. This limitation is eliminated by placing two SCRs back-to-back, as shown in figure 10.15(a), so that each conducts alternately as the polarity reverses. Such a device is called a **triac**, and its symbol is shown in figure 10.15(b). When the triac is connected in a circuit as in figure 10.15(c), its behavior is the same as for the

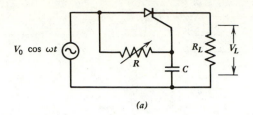

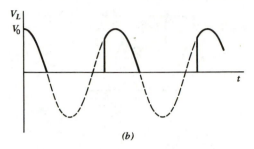

(b)

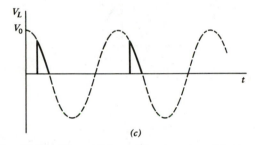

(c)

Fig. 10.14 The addition of the capacitor C in the SCR circuit in (a) allows the output waveform to be varied from nearly a half wave (b) for R small to a narrow spike (c) for R large.

SCR circuit of figure 10.14, except that the output waveform is symmetrical about $V = 0$, as shown in figure 10.15(d).

One difficulty with the gate-trigger circuits shown in figures 10.12–10.15 is the unreliable operation that results from the variation in required trigger currents between different devices and for the same device at different temperatures. A common solution is to insert a neon bulb in series with the gate, so that a large amount of energy is delivered to the gate in a short pulse just after the bulb goes into conduction. For lower voltage operation, a special solid-state trigger diode, called a **diac**, can be used in place of the neon bulb. A diac is like a triac except that it is a two-terminal device that is triggered into conduction when the voltage across its terminals exceeds a prescribed value (typically ~ 30 V) in either direction.

A final example of the many uses of the SCR is the **pulser circuit** shown in figure 10.16(a). In the absence of an input voltage ($V_{in} = 0$), the capacitor C charges

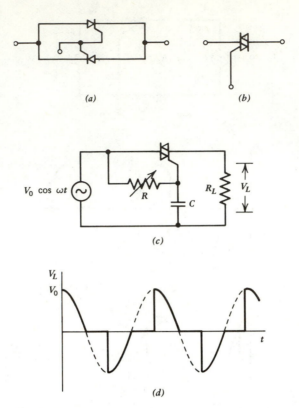

Fig. 10.15 Two SCRs connected as in (a) form a unit called a triac whose symbol is shown in (b). When connected in a circuit as in (c), an output waveform as in (d) results.

to voltage V_0 through resistor R, and no current flows through the SCR. If a positive pulse of sufficient size and duration (usually a few volts for a few microseconds) is applied at V_{in}, as shown in figure 10.16(b), the SCR abruptly goes into conduction and dumps all the energy of capacitor C into the load resistor R_L. The voltage across R_L is thus a decaying exponential

$$V_{out} = V_0 e^{-t/R_L C} \tag{10.4}$$

as shown in figure 10.16(c). When the capacitor voltage drops to a sufficiently low value, the SCR anode current is insufficient to maintain the SCR in conduction (provided R is sufficiently large), and the SCR stops conducting. The capacitor then recharges and awaits the next trigger pulse at V_{in}. A common application for such a circuit would be an electronic photoflash in which the energy stored in a capacitor is used to produce an intense but brief burst of light at a precisely controlled time.

The SCR pulser thus behaves something like a pulse amplifier, in that the output pulse can be much larger than the input pulse. But it differs from an amplifier, in that

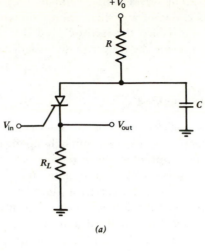

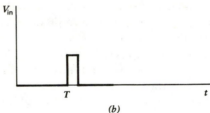

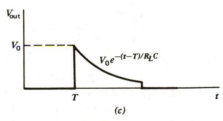

Fig. 10.16 The SCR pulser in (*a*) can be triggered by a small brief pulse (*b*) so as to produce a larger pulse of constant size and shape (*c*).

the shape of the output pulse is determined by V_0, R_L, and C, independent of V_{in} so long as V_{in} is sufficient to trigger the SCR into conduction.

The SCR has two vacuum tube counterparts that are still in common use, especially for very-high-voltage applications. The **thyratron** is a vacuum triode filled usually with low-pressure hydrogen gas and containing an electron-emitting, heated cathode. A positive pulse applied from grid to cathode ionizes the hydrogen and triggers the device into conduction. The **ignitron** is a similar device which is used at even higher voltages and currents (up to ~ 50 kV at 10^6 A). It contains an anode and a pool of mercury that serves as the cathode. The device is triggered into conduction by the application of a positive voltage (usually a few hundred volts) between a third electrode (called the **ignitor**) and the cathode. Thyratrons and SCRs are often used to provide the trigger pulse for ignitrons.

10.6 Optoelectronic Devices

A rapidly developing area of electronics involves devices whose electrical properties are altered by incident light or devices that emit light upon application of a voltage or current. The simplest such **optoelectronic device** is the **photodetector** which consists of a **photoconductor** such as cadmium sulfide whose resistance drops because of the excitation of free charge carriers by incident light. It is a passive device in the sense that it does not produce electrical power. In combination with an external voltage source and ammeter, it can be used to measure light intensity.

The basic semiconductor optoelectronic device is the *pn* junction with a transparent window. In such a **photodiode**, photons incident on the junction cause a flow of current across the junction. When optimized to produce the maximum electrical output power for a given incident light, such devices are called **photovoltaic cells** or **solar cells** and hold promise for large-scale generation of electricity from sunlight. Unfortunately, these devices are expensive and have low efficiency (10–15%), and have found widespread application only in satellites and space vehicles, where sunlight is abundant and other sources of power are prohibitive. On a smaller scale, photocells are used in applications such as reading punched computer cards and tapes and for producing an audio signal from the information coded on movie film. In such applications, the junction is usually reverse-biased, and the reverse current I_0 varies in proportion to the incident light. The *pn* junction is also sensitive to nuclear radiation and has found application in nuclear particle detectors.

An alternate configuration combines a light-sensitive *pn* junction with a bipolar transistor in a device called a **phototransistor**. It results in a multiplication of the light-induced current by the beta of the transistor. The process can be carried one step further in the **photo Darlington transistor** (see figure 8.10). Similarly, an FET can be made sensitive to light. Such a **PHOTOFET** is useful for measuring the attenuation of light passing through liquids and gases, as might be required in a smoke detector.

A photosensitive *pn* junction can also be incorporated into an SCR so that it can be triggered into conduction by a pulse of light. Such a device is called a **light-activated silicon-controlled rectifier** (LASCR). A typical application would be in a remote photographic flash unit that triggers on the light from the flash attached to a camera so as to provide extra illumination on the subject being photographed.

The process that is the reverse of the photodiode is also observed to occur. In the **light-emitting diode** (LED), a forward current through a *pn* junction can be used to produce light. Such junctions are usually made of gallium arsenide or gallium phosphide and typically emit red light, although other colored LEDs are also available. The prime advantages of the LED over other types of light sources are their compact size, low current and voltage requirements, high efficiency, and exceptionally long operating life. For these reasons, LEDs find widespread use in devices such as pocket calculators where battery drain and operating life are prime considerations.

LEDs can also be combined with some form of light-sensitive detector (LSD) such as a photodiode to form a device called an **optocoupler** or **optoisolator**. It produces an output current approximately proportional to the input current but offers a high degree of electrical isolation and voltage standoff between the input and the output, because the emitter and detector can be separated some distance by a transparent insulator or by a grounded, conducting screen. Optoisolators normally operate in the near infrared rather than in the visible portion of the spectrum, and typically provide $>10^{11}$ Ω of isolation.

Often the light emitter and detector are separated by a considerable distance and are coupled by means of a thin flexible **light guide** which often consists of a bundle of thin fibers of glass or plastic (called **fiber optics**). Because the frequency of visible and infrared light is much higher than the frequencies normally encountered in electronic circuits, fiber optics offer the potential of a much higher rate of information transfer than ordinary electrical transmission lines. Fiber optics are thus finding application in telephone, television, and data-transmission systems.

An entirely different type of optoelectronic device is the **liquid crystal display** (LCD), in which an applied voltage changes the opacity of the crystal. When viewed against a dark background, it thus gives the appearance of changing from light to dark. Since such a device does not directly emit light, its power consumption can be extremely small. Such a device would be a natural choice for an application in which the ambient light level is high and available power is minimal, such as a digital wristwatch.

When extremely high sensitivity to light is required, a device called a **photo-multiplier** (**PM**) **tube** is often used. As shown in figure 10.17, such a device is a vacuum tube consisting of a cold, photosensitive cathode that emits electrons when struck by light. The electrons are attracted to a nearby positive electrode (called a **dynode**) where they typically release 3 to 6 secondary electrons which are attracted to the next, even more positive dynode, and the whole process repeats through many stages, producing a large current at the anode. Such a device is so sensitive that it can detect a single photon of light, in which case the output consists of a negative voltage pulse with a size determined by the capacitance of the measuring instrument.

An alternate and more compact configuration is the **channeltron electron multiplier** (CEM) in which a long, thin, evacuated glass tube is coated on the inside with a low-work-function, conducting material that takes the place of the individual dynodes in the photomultiplier tube. A high voltage is applied between the ends of the tube, and an avalanche of electrons is formed whenever light is incident on the more negative end of the coating. Arrays of such devices can be used to form a two-dimensional electronic image of the object being viewed.

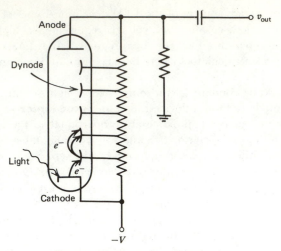

Fig. 10.17 Photomultiplier tube.

10.7 Summary

This chapter has dealt with a small number of the more common nonlinear circuits and devices. Oscillators are circuits that do the opposite of rectifiers. They convert a dc voltage into an ac voltage. There are dozens of ways to make oscillators. They all involve either active circuit components with positive feedback or a device with negative resistance. The frequency of an oscillator is often determined by a resonant LC circuit, but RC circuits and quartz crystals can also be used. Oscillators can be made with sinusoidal outputs or with a variety of other periodic waveforms.

The multivibrator is a circuit consisting of a pair of transistors that abruptly switch between saturation and cutoff. They can be made in three configurations. The bistable multivibrator is equally happy in either of two states. The monostable multivibrator can be made to change states, but it always returns to its initial state after a prescribed interval. The astable multivibrator isn't really happy in either state and continually switches back and forth. It is just another kind of oscillator. The Schmitt trigger is a variation of the monostable multivibrator that resembles the op-amp comparator.

The list of semiconductor devices in widespread use is a very long one. Tunnel diodes are useful as oscillators and high-speed switching devices because of their negative resistance characteristics. The unijunction transistor is a three-terminal device with a negative resistance characteristic. It and the silicon controlled rectifier are switching devices, but the means by which they are turned off once they begin to conduct are rather different. Optoelectronic devices are electronic devices that either respond to or emit light. Their uses are numerous and rapidly growing.

Problems

10.1   Assume the op amp in figure 10.1(a) is real with $A_0 = 10^4$ and $r_{in} = 1\ k\Omega$. Calculate the maximum value of R for which oscillation will occur.

10.2   Show that the frequency of oscillation of the circuit below is given by $f_0 = 1/2\pi\sqrt{6}\ RC$ provided the op amp output does not saturate.

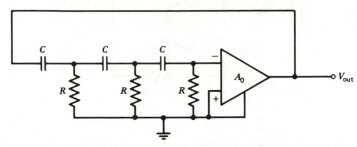

10.3   For the circuit in problem 10.2, show that oscillation will occur only if the open loop voltage gain of the op amp exceeds $A_0 = 29$.

10.4   For the quartz crystal equivalent circuit shown in figure 10.4(b), assume $R = 0$ and $C_2 \gg C_1$. Show that the circuit exhibits a series resonance at an angular frequency ω_0 that is independent of C_2 and that a parallel resonance occurs at a slightly higher frequency such that the difference between the two resonances is given by $\Delta\omega = \omega_0 C_1/2C_2$.

10.5   If the crystal described in figure 10.4(b) with $R = 1000\ \Omega$ is used in the oscillator circuit of figure 10.5 with $R = 1000\ \Omega$, what is the minimum ratio R_f/R_i for which oscillation will occur?

10.6   Calculate the values of L and C_1 for the crystal equivalent circuit of figure 10.4(b) if the crystal has a series resonance at 1 MHz with a Q of 10^5 and a series resistance of $R = 100\ \Omega$.

10.7   Calculate the minimum value of V_{in} which will trigger the Schmitt trigger circuit in figure 10.7, assuming the transistors are ideal, $R_1 = 2\ k\Omega$, $R_2 = R_3 = R_4 = R_E = 1\ k\Omega$, and $V = 24\ V$.

10.8   Show that the Schmitt trigger circuit described in problem 10.7 exhibits hysteresis; that is, once the circuit is triggered, show that V_{in} must fall to a value below the value that caused it to trigger before it switches back to its initial state, and calculate that value of V_{in}.

10.9   Design a monostable multivibrator using the tunnel diode whose characteristics are shown in figure 10.8(a). The circuit should produce a pulse of duration $\sim 10^{-3}$ s.

10.10   Estimate the minimum value of resistance that could be placed in parallel with the inductor in figure 10.8(b) and still allow the circuit to oscillate if the tunnel diode is described by the curve in figure 10.8(a). What value should V_0 have for this value of resistance?

10.11   The circuit below is an astable multivibrator that uses an ideal op amp with an output saturation voltage of ± 10 V. Calculate the period of the output square wave.

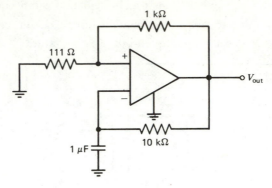

10.12   Sketch the voltage waveform across the capacitor in the circuit in problem 10.11 and indicate the magnitude of the voltage.

10.13   Solve the homogeneous differential equation for the voltage across a parallel RLC in which the resistance is negative.

10.14   Calculate the frequency of the neon bulb relaxation oscillator of figure 10.9 in terms of RC, assuming a power-supply voltage of $V_0 = 160$ V.

10.15   Derive a criterion for the values of β for the two transistors in figure 10.11(c) that will ensure that the circuit behaves like an SCR.

10.16   Calculate the average power dissipated in the load R_L in figure 10.12 (a) assuming the voltage source is sinusoidal with peak value V_0 and the SCR is triggered whenever the gate current exceeds I_t.

10.17   Calculate the dc and fundamental frequency component for the Fourier series for the waveform in figure 10.12(b) if the SCR conducts for a quarter-cycle with a peak output voltage of $V_0 = 100$ V.

10.18   Design a circuit using an SCR that can be triggered with a $+10$-V, 1-μs-wide voltage pulse which will produce an exponentially decaying pulse of $+100$-V initial amplitude and 100-μs time constant across a 100-Ω load. Assume that a gate current of 1 mA will reliably trigger the SCR and that the holding current is also 1 mA.

10.19   If an SCR pulser is triggered repetitively at too high a repetition rate, the capacitor will not charge up to its full voltage. Derive an expression for the peak output voltage as a function of the repetition rate f (in hertz) for the conditions described in problem 10.18. (Assume $f \ll 1/R_L C$.)

10.20   How many dynodes would be required for a photomultiplier tube to produce a >0.1-V pulse across a 100-pF load when a single electron is emitted from the photocathode if each dynode emits 5 electrons per incident electron?

chapter 11

Digital
Circuits

11.1 Binary Numbers

In most of the applications of electronic circuits encountered so far, the voltages and currents vary continuously over a range of values. Such circuits are called **analog circuits**. As the opposite extreme are circuits in which the voltages and currents are allowed to have only two rather different values, as was the case with the multivibrators described in section 10.2. Such circuits are called **digital circuits**. With digital circuits, the exact values of the voltages are of no consequence so long as one can unambiguously determine which of the two states the circuit is in. The two states are variously referred to as on/off, true/false, yes/no, high/low, or one/zero. Digital circuits are inherently more reliable and less prone to noise and interference than analog circuits.

Digital circuits lend themselves rather naturally to performing arithmetic with **binary** numbers. A binary number is a number composed of the two binary digits (called **bits**), 0 and 1. With a decimal number, such as 931, the decimal digits represent successive powers of ten:

$$931_{10} = 1 \times 10^0 + 3 \times 10^1 + 9 \times 10^2$$

Similarly, a binary number such as 10110 can be expressed as successive powers of two:

$$10110_2 = 0 \times 2^0 + 1 \times 2^1 + 1 \times 2^2 + 0 \times 2^3 + 1 \times 2^4 = 22_{10}$$

The idea can be generalized to numbers of any base. **Octal** (base 8) and **hexadecimal** (base 16) representations are quite common. The binary, octal, and hexadecimal equivalents of the decimal numbers 1 through 20 are given in table 11.1. Each octal digit can be represented as a three-bit number, and each hexadecimal digit can be represented as a four-bit number. An eight-bit binary number is often called a **byte**. A byte can thus be written as a two-digit hexadecimal number.

Binary numbers are added just like decimal numbers, by carrying a digit

TABLE 11.1   Representations of Decimal Numbers in Binary, Octal, and Hexadecimal

Decimal	Binary	Octal	Hexadecimal
1	1	1	1
2	10	2	2
3	11	3	3
4	100	4	4
5	101	5	5
6	110	6	6
7	111	7	7
8	1000	10	8
9	1001	11	9
10	1010	12	A
11	1011	13	B
12	1100	14	C
13	1101	15	D
14	1110	16	E
15	1111	17	F
16	10000	20	10
17	10001	21	11
18	10010	22	12
19	10011	23	13
20	10100	24	14

whenever the result would be more than one digit:

$$\begin{array}{r} 110 \\ + 101 \\ \hline 1011 \end{array}$$

Similarly, two binary numbers can be multiplied as with decimal numbers:

$$\begin{array}{r} 10110 \\ \times \quad 101 \\ \hline 10110 \\ 00000 \\ 10110 \\ \hline 1101110 \end{array}$$

Note that the multiplication of two binary numbers consists merely of successive additions of a column of numbers formed by shifting the bits of the original number one place to the left for each bit in the multiplier that is a 1.

Subtraction of binary numbers follows the same rules as subtraction of decimal numbers. It is necessary to borrow from the next higher bit whenever one tries to subtract a 1 from a 0. A convenient trick for the subtraction of two binary numbers makes use of what is called **two's complement arithmetic**. The rule is to take the number that is to be subtracted, and invert it (change all 0's to 1 and all 1's to 0), then add 1, and add the result to the original number, discarding any leftover carry bit. For example, to subtract $43_{10} = 101011_2$ from $57_{10} = 111001_2$, proceed as follows:

$$
\begin{array}{r}
111001 \\
+\,010100 \\
+\qquad 1 \\
\hline
1001110
\end{array}
$$

The leftmost bit of the result is a carry bit and is discarded, leaving the number $001110_2 = 14_{10}$. In this notation the remaining leftmost bit (0 in this case) is called the **sign bit**. Its only purpose is to indicate whether the number is positive or negative. A 0 sign bit means that the remaining number is positive; a 1 sign bit means that it is negative. If the result is negative, it is necessary to take its inverse and add 1 to find its value.

The division of two binary numbers can be done by counting how many times one number can be subtracted from the other. The count then becomes the quotient. Note that unlike the other arithmetic operations, the division of one integer by another does not, in general, give an integer result. The remainder after subtracting one number from the other as many times as possible then can be used to determine the fractional part of the quotient.

An important conclusion that can be drawn from this discussion is that all arithmetic operations on binary numbers can be reduced to addition. Thus a digital circuit capable of performing binary addition becomes the building block for all the more complicated operations.

11.2 Logic Gates

Electronic circuits that perform operations with binary digits are called **logic gates**. For example, a circuit with two inputs and one output that produces a 1 at its output if input A and input B are both 1, and 0 otherwise, is called an **AND** gate. A circuit with such a property is shown in figure 11.1(a). In this circuit, if either A or B is at zero V, one of the diodes (assumed ideal) is forward biased, and the output V_0 is zero. If, on the other hand, both A and B are at $+V$ (typically 5 V), the output V_0 is at $+V$, which we identify as a binary 1. The behavior of such circuits is described by a **truth table**, as shown in figure 11.1(b). A truth table describes the output for all possible combinations of input variables. Examination of the truth table shows that the **AND** gate is essentially a binary multiplication circuit, and so the **AND** operation is indicated by a dot:

$$V_0 = A \cdot B$$

Sometimes the dot is omitted and the **AND** operation is written simply as

$$V_0 = AB$$

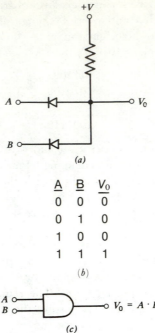

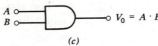

A	B	V_0
0	0	0
0	1	0
1	0	0
1	1	1

(b)

$V_0 = A \cdot B$

(c)

Fig. 11.1 (a) **AND** circuit. (b) Truth table. (c) Symbol.

but in this text the dot will always be used. The symbol for the **AND** gate is shown in figure 11.1(c). An **AND** gate can have more than two inputs, in which case it produces a 1 at its output only if all the inputs are 1. An **AND** gate with more than two inputs can be used as a two-input **AND** gate if the unused inputs are maintained at 1. The **AND** gate is also called a **coincidence circuit**, since it produces an output if two positive pulses at the inputs coincide in time.

A different type of circuit produces a 1 at its output if A or B (or both) are 1; and 0 otherwise. It is called an **OR** gate. Such a circuit is shown in figure 11.2(a). Its operation is described by the truth table in figure 11.2(b). The **OR** operation is indicated by a $+$ sign:

$$V_0 = A + B$$

The symbol for the **OR** gate is shown in figure 11.2(c). An **OR** gate can have more than two inputs, in which case it produces a 1 at its output if any of the inputs are 1. An **OR** gate with more than two inputs can be used as a two-input **OR** gate if the unused inputs are maintained at 0.

One limitation of the diode logic gates described above is the degradation of the signal (~ 0.6 V/gate) that occurs when a series of such gates are connected together. Furthermore, the number of gates that can be connected to the output of a diode gate (called the **fanout**) is limited to a fairly small number, since each additional gate

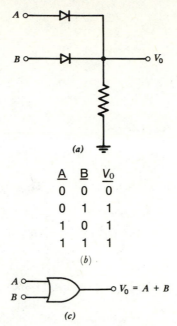

(a)

A	B	Vo
0	0	0
0	1	1
1	0	1
1	1	1

(b)

$V_0 = A + B$

(c)

Fig. 11.2 (a) OR circuit. (b) Truth table. (c) Symbol.

loads down the voltage output to the point where eventually the 1 state is no longer reliably recognized. Consequently, logic gates are more often constructed with transistors, as shown in figure 11.3(a). Such a circuit resembles the OR gate, except that the transistor inverts the output as shown by the truth table in figure 11.3(b). Such a circuit is called a **NOR** gate (**NOT-OR**). The **NOT** operation is indicated by a

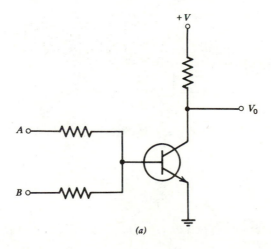

$+V$

V_0

(a)

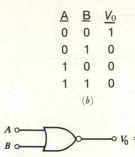

A	B	V_0
0	0	1
0	1	0
1	0	0
1	1	0

(b)

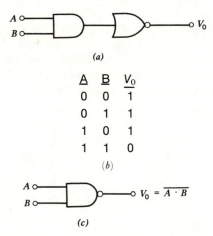

$V_0 = \overline{A + B}$

(c)

Fig. 11.3 (a) NOR circuit. (b) Truth table. (c) Symbol.

bar over the quantity, so that the **NOR** operation is indicated by

$$V_0 = \overline{A + B}$$

The symbol for the **NOR** gate is shown in figure 11.3(c). A **NOR** gate can have any number of inputs. A single input **NOR** gate or a multiple input **NOR** gate with all but one of its inputs maintained at 0 performs the **NOT** operation (inversion).

Finally, one can produce a **NAND** gate (**NOT-AND**) by combining an **AND** gate with a single input **NOR** gate, as shown in figure 11.4(a). The **NAND** operation is

A ○———⊐D⊐—⊐D○——○ V_0
B ○

(a)

A	B	V_0
0	0	1
0	1	1
1	0	1
1	1	0

(b)

A ○
B ○———⊐D○——○ $V_0 = \overline{A \cdot B}$

(c)

Fig. 11.4 (a) NAND circuit. (b) Truth table. (c) Symbol.

indicated in the truth table in figure 11.4(b), and written as

$$V_0 = \overline{A \cdot B}$$

The symbol for the **NAND** gate is shown in figure 11.4(c). A **NAND** gate can have any number of inputs. Any unused inputs of a **NAND** gate must be held at 1 or connected in parallel with one of the inputs that is being used. A single input **NAND** gate or a multiple input **NAND** gate with all but one of its inputs maintained at 1 performs the **NOT** operation. The **NOT** operation is sometimes denoted by the symbol in

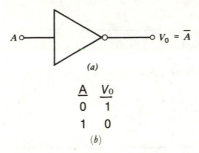

$$A \circ\!\!\!-\!\!\!\!-\!\!\!\!\triangleright\!\!\!\!-\!\!\!\!-\!\!\!\circ V_0 = \overline{A}$$

(a)

A	V_0
0	1
1	0

(b)

Fig. 11.5 **NOT** circuit. (a) Symbol. (b) Truth table.

figure 11.5(a) in which the resemblance to an inverting op amp is more than coincidental. The truth for the **NOT** operation is given in figure 11.5(b). Notice that a single input **NOR** gate, a single input **NAND** gate, and a **NOT** gate all perform the same function.

Not surprisingly, logic gates are available as integrated circuits. Such circuits are often more complicated than those shown in figures 11.1–11.4 in order to improve reliability, reduce power dissipation, and increase operating speed. Speed and power dissipation can usually be traded off against one another. This is an important fact, since large digital circuits such as computers typically employ thousands of logic gates, and the speed and total power dissipation are usually limiting factors.

Because of the limitations previously mentioned, gates containing only **diode and resistor logic** (DRL) components are rarely used. Some of these limitations are overcome with **resistor-transistor logic** (RTL), such as was shown in figure 11.3(a). Some additional improvements, especially in physical size, noise immunity, and fanout, can be achieved by a combination of diodes and transistors (DTL). The diodes, however, have considerable junction capacitance, and thus the speed with which the gate can operate is not great. Faster operation is achieved with **transistor-transistor logic** (TTL) in which special transistors with multiple emitters are used. Even greater speed is achieved by using **emitter-coupled logic** (ECL) in which emitter followers are used for low output resistance, and the input capacitance is reduced by not driving the transistors into saturation. ECL gates are the most expensive and have the highest power requirements, but they have a large fanout and are able to respond in about 10^{-9} s. At the opposite extreme is the MOS logic family which uses MOSFET devices rather than bipolar transistors. Because of the very high input resistance of the MOSFET, the power dissipation per gate is very small and the fanout enormous, but the switching speed is comparatively slow. Finally, one should note that the voltage levels used for the different types of logic are not necessarily the same, and so one should avoid mixing types unless care is taken to convert the voltages properly. The characteristics of a typical logic gate employing the various logic types are summarized in table 11.2.

TABLE 11.2 Comparison of a Typical Logic Gate Employing Various Logic Types

Logic Type	Speed	Power	$V_0(0/1)$	Fanout
RTL	50 ns	10 mW	0.2/0.9	4
DTL	25 ns	15 mW	0.2/4.0	8
TTL	10 ns	20 mW	0.2/3.0	10
ECL	2 ns	50 mW	−1.55/−0.75	24
MOS	200 ns	0.3 mW	0/3–15	50

11.3 Boolean Algebra

The analysis of logic networks is facilitated by a set of theorems developed by George Boole, an English mathematician, and known as **Boolean algebra**. These theorems were developed for symbolic logic long before the advent of digital electronics. Most of the theorems of Boolean algebra are identical to those of ordinary algebra:

$$\text{Commutation:} \quad \left.\begin{array}{l} A + B = B + A \\ A \cdot B = B \cdot A \end{array}\right\} \tag{11.1}$$

$$\text{Association:} \quad \left.\begin{array}{l} (A + B) + C = A + (B + C) \\ (A \cdot B) \cdot C = A \cdot (B \cdot C) \end{array}\right\} \tag{11.2}$$

$$\text{Distribution:} \quad A \cdot (B + C) = (A \cdot B) + (A \cdot C) \tag{11.3}$$

Certain other theorems arise from the binary properties of the quantities:

$$A \cdot A = A \tag{11.4}$$

$$A + A = A \tag{11.5}$$

$$A \cdot \bar{A} = 0 \tag{11.6}$$

$$A + \bar{A} = 1 \tag{11.7}$$

$$\bar{\bar{A}} = A \tag{11.8}$$

A very useful relationship of Boolean algebra is **De Morgan's theorem**:

$$\overline{A + B} = \bar{A} \cdot \bar{B} \tag{11.9}$$

By replacing A with $\bar{A}$ and B with $\bar{B}$ in equation (11.9), and using equation (11.8), an alternate form of De Morgan's theorem can be derived:

$$\bar{A} + \bar{B} = \overline{A \cdot B} \tag{11.10}$$

De Morgan's theorem is very important, because it allows all the basic logic operations to be done with only **NOR** gates or with only **NAND** gates. For example, figure 11.6(a) shows how an **AND** gate can be constructed from **NOR** gates, and

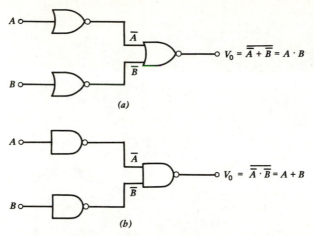

(a)

(b)

Fig. 11.6 (a) Construction of an **AND** gate using **NOR** gates. (b) Construction of an **OR** gate using **NAND** gates.

figure 11.6(b) shows how an **OR** gate can be constructed from **NAND** gates. Many other relationships can be derived from these theorems.

All these theorems can be proved quite simply by writing out the truth tables. For example, the truth table for De Morgan's theorem is given in table 11.3.

TABLE 11.3   Truth Table for Proving De Morgan's Theorem

A	B	$\bar{A}$	$\bar{B}$	$\bar{A} \cdot \bar{B}$	$A + B$	$\overline{A + B}$
0	0	1	1	1	0	1
0	1	1	0	0	1	0
1	0	0	1	0	1	0
1	1	0	0	0	1	0

11.4   Logic-Gate Applications

Logic gates can be combined to perform a variety of mathematical operations. For example, the circuit in figure 11.7(a) will add two single-bit numbers, A and B, and produce a two-bit number with right- and left-hand digits, R and L, respectively. Such a circuit is called a **half-adder circuit**, and it is described by the truth table in figure 11.7(b). The half-adder circuit has limited usefulness, because the most complicated problem it can solve is $1 + 1 = 2$. To add larger numbers (many bits), it

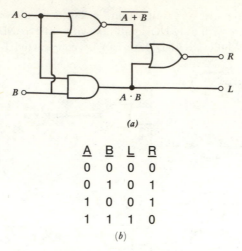

(a)

A	B	L	R
0	0	0	0
0	1	0	1
1	0	0	1
1	1	1	0

(b)

Fig. 11.7 The half-adder circuit in *(a)* is described by the truth table in *(b)*.

is necessary to have an adder circuit in which the carry bit from the preceding circuit can be added to the right-hand digit. The left-hand digit then becomes the carry bit for the next adder, and so forth. Such a circuit is called a **full-adder**, and is shown in figure 11.8(a). The operation of the full-adder is described by the truth table in figure 11.8(b). A full-adder is capable of adding two single-bit numbers (A and B) plus a carry bit (C) and producing a single-bit result (R) and a carry bit (L).

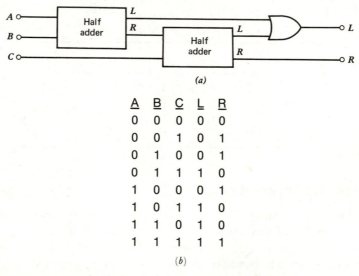

(a)

A	B	C	L	R
0	0	0	0	0
0	0	1	0	1
0	1	0	0	1
0	1	1	1	0
1	0	0	0	1
1	0	1	1	0
1	1	0	1	0
1	1	1	1	1

(b)

Fig. 11.8 The full-adder circuit in *(a)* is described by the truth table in *(b)*.

With full-adders, one can proceed to construct circuits to perform nontrivial calculations. For example, suppose one wishes to add the binary numbers $1101 = 13_{10}$ and $1001 = 9_{10}$. For the first number, one would set four voltages as appropriate, beginning with the right-hand (least-significant) bit:

$$A_1 = 1 \qquad A_2 = 0 \qquad A_3 = 1 \qquad A_4 = 1$$

Similarly for the second number:

$$B_1 = 1 \qquad B_2 = 0 \qquad B_3 = 0 \qquad B_4 = 1$$

These eight voltages could then be applied simultaneously to the circuit in figure 11.9(a), and the output C would be a five-bit number $10110 = 22_{10}$. Such a circuit is called a **four-bit adder**, and it can add numbers up to $15 + 15 = 30$. Clearly, by stacking four-bit adders together, numbers of any size can be added. The addition of

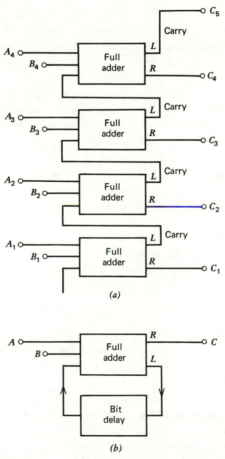

(a)

(b)

Fig. 11.9 (a) Four-bit parallel adder. (b) Serial adder.

2 eight-digit decimal numbers, such as might be done with an electronic calculator, requires 7 four-bit adders.

The four-bit adder just described is an example of a **parallel operation**, in that all of the input voltages must appear simultaneously, and a separate connection is required for each bit of each number being added and for each bit of the output. An alternate technique is to use a **serial operation**, in which each bit of the numbers is transmitted sequentially, beginning with the least-significant bit. Such a circuit for the addition of two numbers is shown in figure 11.9(b). Suppose at $t = 0$, the right-hand digits of the numbers 1101 and 1001 are applied at A and B, respectively. The output at C would be 0 and at L would be 1. Now suppose at $t = 1$ μs, A and B are changed to 0 and 0 to represent the next digit of each number, and the bit delay circuit provides a 1 carry-bit that was generated 1 μs earlier. The bit delay circuit could be a monostable multivibrator. The output at 0 would then be 1 and at L would be 0. The process would continue, producing a train of pulses at C representing successively the bits of the result. With serial addition, there is no limit to the size of the numbers that can be added. Furthermore, serial addition requires fewer components than parallel addition. However, serial addition is much slower, because the bits have to be fed in one at time rather than all at once.

Where data has to be transmitted over long distances, serial transmission is invariably used, since this permits the use of a single two-conductor transmission line. The rate of transmission is expressed in terms of bits per second, called the **baud rate**. It should be clear from the discussion of Fourier analysis in Chapter 5 that the maximum baud rate is proportional to the bandwidth of the circuit, and that a bandwidth of $\gtrsim 10$ kHz is required to transmit 10,000 bits per second.

Much digital data transmission takes place over the already existing telephone network. For this purpose, a device called a **modem** (modulation-demodulation) is used. It generates a 2225-Hz tone whenever a logical 1 is to be sent and a 2025-Hz tone whenever a logical 0 is to be sent. A second pair of frequencies is used for reception, with a logical 1 at 1270 Hz and a logical 0 at 1070 Hz. Such a low-speed modem might operate typically at 300 baud.

The adder circuits described above can be combined with other logic gates to perform additional operations such as multiplication, subtraction, and division, using the rules described in section 11.1. Circuits are also made that perform special operations such as the evaluation of transcendental functions (sine, cosine, log, etc.), direct conversions (inches to meters, °C to °F, etc.) and binary to decimal conversions (and vice versa). Such circuits are called **read-only memories** (ROMs), **translators**, or **decoders**. They are available singly or in combination as integrated circuits at very nominal cost.

Although ROMs are normally programmed during manufacture to perform a definite operation, **programmable read-only memories** (PROMs) are also available, in which the user can specify the operation to be performed by the application of appropriate voltages that permanently alter the internal circuitry of the device. Some types of PROMs are erasable, so that they can be reused for different purposes or corrected if a mistake is made during the programming. With

EPROMs (erasable PROM), the device must be removed from the circuit and exposed to ultraviolet radiation to erase its contents. With EAROMs (electrically alterable ROM), the contents of the device can be selectively altered by the application of appropriate voltages while it is still in the circuit.

11.5 Flip-Flops

Flip-flop circuits similar to the op-amp latch and the bistable multivibrator can also be constructed from logic gates. For example, figure 11.10(a) shows a circuit called an RS flip-flop in which two **NOR** gates are used to produce an output at Q which is

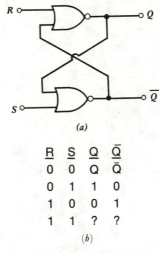

(a)

R	S	Q	$\bar{Q}$
0	0	Q	$\bar{Q}$
0	1	1	0
1	0	0	1
1	1	?	?

(b)

Fig. 11.10 The RS flip-flop in (a) is described by the truth table in (b).

either 0 or 1. The inverse quantity will appear at $\bar{Q}$. If $S=1$ and $R=0$, Q will go to 1 and remain there until R is set to 1 and S to 0, which resets the circuit and causes Q to go to zero. The truth table for the RS flip-flop is shown in figure 11.10(b). A serious drawback of the RS flip-flop appears if both inputs are set to 1. In such a case Q and $\bar{Q}$ both go to 0, and the final state after the inputs are returned to 0 will have a value that depends on asymmetries in the circuit or on which input first goes to 1.

A variation of the RS flip-flop shown in figure 11.11(a) has a third input, called the **clock** input. The circuit ignores any signals at the R and S inputs unless the clock input is at 1. The clock input allows one to control the time at which the circuit switches state. The truth table for the clocked RS flip-flop is the same as the truth table for the simple RS flip-flop in figure 11.10.

Another type of flip-flop is the D (for **data**) flip-flop shown in figure 11.11(b). It has an R, S, and clock input and also a fourth input, called the D input. The R and S are used to put the circuit in one of two possible initial states. When the clock is set to

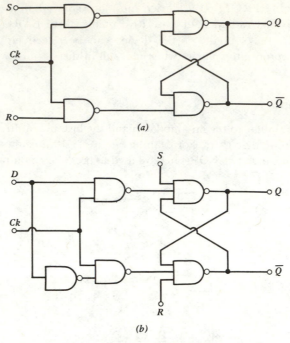

Fig. 11.11 (a) Clocked RS flip-flop. (b) D flip-flop.

1, the output Q switches to 0 if D is 0 or 1 if D is 1, and remains there until reset or until another clock pulse arrives.

The most complicated and most useful flip-flop is the JK flip-flop. It has two inputs, J and K, and two outputs, Q and $\bar{Q}$, as shown in figure 11.12(a). The circuit is caused to switch state by a momentary 1 pulse at the clock (Ck) input. The term "toggle" is sometimes used instead of "clock." The circuit remembers the values of J and K at the instant the clock goes from 0 to 1, but it does not switch the output state until the clock input goes back to 0. The behavior of the JK flip-flop is described by the truth table in figure 11.12(b), in which the subscript i (initial) refers to the value of the quantity when Ck goes from 0 to 1, and f (final) refers to the value when Ck goes back to zero. The circuit is reset by means of the Cl (clear) input. Normally Cl is kept at 1. Setting Cl to 0 sets Q to 0 and $\bar{Q}$ to 1. The JK flip-flop described here is actually two flip-flops and is an example of a **master-slave flip-flop**. The master flip-flop is set by the input (J and K) when the clock goes to 1. It then commands the slave to produce the appropriate output (Q and $\bar{Q}$) when the clock goes back to 0.

JK flip-flops are useful in devices such as multivibrators, counters, and shift-registers. Multivibrators have been discussed earlier (section 10.2). A **counter** is a device that produces a digital output equal to the number of pulses that have appeared at the input since the device was last cleared. Figure 11.13(a) shows how such flip-flops are connected to make a binary counter. A common application of a

268 **Digital Circuits**

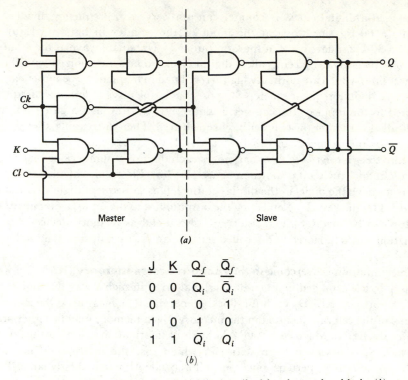

J	K	Q_f	$\bar{Q}_f$
0	0	Q_i	$\bar{Q}_i$
0	1	0	1
1	0	1	0
1	1	$\bar{Q}_i$	Q_i

(b)

Fig. 11.12 The JK flip-flop in (*a*) is described by the truth table in (*b*).

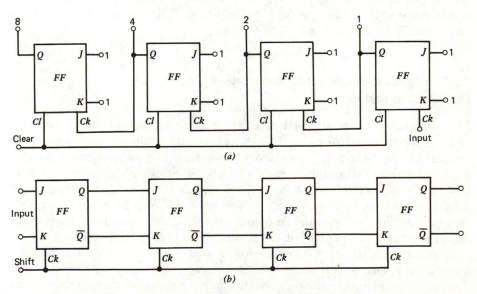

Fig. 11.13 (*a*) Binary counter. (*b*) Shift-register.

counter is in a digital clock. A counter can also serve as a frequency divider. For example, a 60-Hz sine wave at the input of the counter in figure 11.13(a) will produce a 30-Hz square wave at the 2 output, a 15-Hz square wave at the 4 output, and so on. Such divider circuits allow digital clocks to be synchronized with the 60-Hz power lines, which are normally highly regulated in frequency. A counter can also be used as a digital frequency meter. Suppose that a sine wave of unknown frequency is applied to the input of the counter in figure 11.13(a) through a gate that opens periodically for 1 s after the counter has been cleared. Then the reading after the gate closes will be the frequency of the sine wave in Hz (expressed as a binary number).

A **shift register** is a device that moves each digit of a number one place to the right or left. Such a circuit is shown in figure 11.13(b). Electronic calculators use shift registers to move the digits in the display to the left as subsequent digits of a number are entered by the keys. By connecting the output back to the input, a **circular shift register** can be made. Shift registers are also useful as memory devices, but the information has to be stored and recalled serially, and so such devices tend to be very slow.

This limitation is overcome in the **random-access memory** (RAM) in which the flip-flops are arranged in a two-dimensional array in such a way that one can set (WRITE) or test (READ) each flip-flop by energizing or measuring the **x-** and **y-address** of the relevant location in the memory. Such memory may be either **static** (or **nonvolatile**), in which case the RAM holds its information so long as the power is not removed, or **dynamic** (**volatile**), in which case the information has to be periodically refreshed, perhaps once per millisecond. Although the dynamic RAM requires more circuitry, it is often preferred because of its lower power consumption and because the use of integrated MOSFET circuitry allows extreme miniaturization. A dynamic RAM capable of storing $2^{10} = 1024$ bytes (1 K) of information need be no larger than a medium size individual transistor.

11.6 Digital-to-Analog and Analog-to-Digital Conversion

In electrical circuits it is often necessary to convert from a digital to an analog signal and vice versa. Digital-to-analog (D-to-A) conversion is relatively easy, as indicated in figure 11.14(a) where a four-bit binary number is converted to an analog voltage V_{out} by an operational amplifier. Such a circuit can be extended to any number of bits. One must only be careful that the op amp does not saturate for the largest possible number at the input. The feedback resistor R_f can be adjusted to give a conveniently large but unsaturated output voltage. If the inherent accuracy of the digital number is to be preserved, the resistors have to be of high tolerance.

Analog-to-digital (A-to-D) conversion is somewhat more complicated, and a wide variety of such circuits has been developed. A common feature of these circuits is that they all involve an oscillator, a counter, and a comparator. Perhaps the simplest such circuit is the one shown in figure 11.14(b). A highly stable oscillator (called a

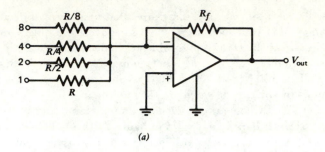

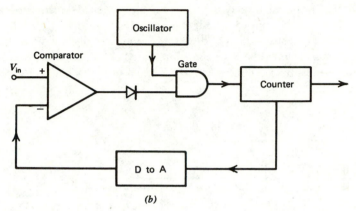

Fig. 11.14 (*a*) Digital-to-analog converter. (*b*) Analog-to-digital converter.

clock) is used. The clock output is fed through an **AND** gate into a counter that counts the number of clock pulses after being reset. The counter produces a digital output which can be converted to an analog signal with a D-to-A converter as described above. It is then compared with the analog input by an analog comparator. When the reading of the counter equals the input voltage V_{in}, the gate closes, and the counter stops and waits to be reset. A common application of an A-to-D converter is a digital voltmeter (**DVM**). In a similar fashion, any quantity that can be represented by a voltage can be converted to a digital format for display or for storing in a digital computer. A-to-D conversion is inherently slow, especially when the counter must be reset to zero for every measurement, as was the case for the circuit just described. More elaborate circuits partially alleviate this difficulty, but speed is still a serious limitation in the use of A-to-D converters for many applications, especially when high resolution is also required.

For A-to-D converters as well as most other digital devices, it is necessary eventually to display the numerical result for interpretation by a human being. Perhaps the simplest such display would be a series of lamps that would light for binary 1 and remain off for a binary 0. The binary number could then be read directly. Unfortunately, the decimal number system is deeply ingrained in our thinking, and so for most purposes it is imperative that some form of decimal display be provided. The first step in such a process is to convert the binary number to a **binary-coded decimal** (BCD) format in which each decimal digit is represented by a four-bit binary number. Thus the binary number $1001110 = 78_{10}$ could be represented in BCD as 0111 1000. Other BCD codings are also possible. Then each four-bit segment of the BCD code would be fed into the input of a decoder logic, as shown in figure 11.15(a) whose output would illuminate appropriate segments of a **seven-segment digital display** as shown in figure 11.15(b). The display could be LEDs or an LCD. A separate display with its own decoder driver would be required for each decimal digit of the display. Alternately, a single decoder driver could be rapidly switched from one display to the next in a process called **multiplexing**. This technique also conserves power, and if the switching is sufficiently rapid, no flicker will be noticed. The construction of decimal readout circuits is tedious, but, as with most other digital circuits, the marvels of integrated circuit technology coupled with the economy of mass production have filled our pockets with such devices for less than the price of a pair of shoes.

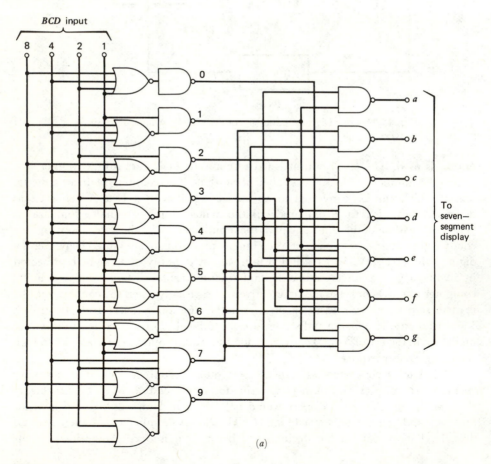

(a)

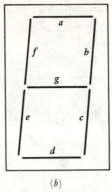

(b)

Fig. 11.15 (a) BCD decoder logic for seven-segment display. (b) Seven-segment digital display.

11.7 Digital Computers

By now, it should be obvious that most any mathematical or logical operation can be performed by a sufficiently complicated arrangement of the basic logic circuits previously discussed. Such **hardwired** circuits have been used in a wide variety of scientific, industrial, and consumer applications. However, it is not very efficient and economical to construct a different circuit for each and every such application. A much better approach is to construct a single device that is capable of performing all the basic operations in any combination and sequence and to provide the user with a means for instructing the device which operations to perform and in what order. Such a device is called a **digital computer**, and it represents the most sophisticated example of digital circuitry. It is outside the scope of this text to provide more than a general introduction to some of the more important aspects of digital computers.

To gain some appreciation of how a circuit can be instructed to perform different operations, consider the circuit in figure 11.16(a). It is called an **exclusive-OR** circuit, because it produces a 1 at its output if either A or B are 1 but not both, as shown by its truth table in figure 11.16(b). But in the present context we can think of it as a circuit that performs an operation on A that is controlled by an instruction at B. If B is set to 0, the operation is to do nothing ($V_0 = A$). If B is set to 1, the operation is to perform an inversion ($V_0 = \bar{A}$). If we now imagine a string of 1's and 0's appearing at A and B in sequence, we can think of A as constituting input data and B as constituting a predetermined **program** of instructions, and the result of the programmed operation on the data would appear as a string of 0's and 1's at the output V_0. Such a device is a primitive and not very useful computer.

Now consider some of the refinements that would be necessary to make a truly useful and versatile device. First, we would need some method for storing the program and possibly the input and output data as well. This **memory** might consist of an array of flip-flops or an array of ferromagnetic toroidal cores that can be

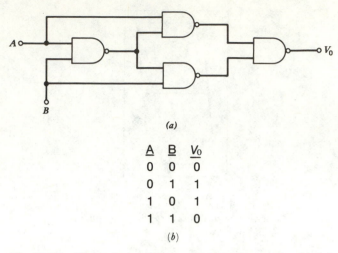

(a)

A	B	V_0
0	0	0
0	1	1
1	0	1
1	1	0

(b)

Fig. 11.16 The exclusive-OR circuit in (a) whose truth table is in (b) is an example of controlling the operation of a circuit by means of an instruction.

magnetized in either of two directions. Information is stored in memory in the form of a series of 0's and 1's called **words**, which are typically 8, 16, or 32 bits long. Some types of information such as the startup instructions for the computer might be stored in a ROM, but to exploit the great versatility of the computer, a significant amount of RAM capability is generally desired. The ability of the computer to store a complicated and easily modified program that may, in fact, specify different operations — depending on the outcome of previous calculations performed on the data — is what makes the computer so versatile, even sometimes giving the impression that the computer has intelligence.

In addition to memory, we would like to provide the computer with a more complete set of instructions to replace the one-bit inversion instruction of the exclusive-OR. We would probably want to implement all the standard arithmetic and logic operations. This will, of course, require a separate multibit number for each instruction that we wish to implement. If we allow our instructions to have eight bits (two bytes), we can choose among 256 possible operations by setting the proper voltages on the eight control lines. Such a device is called an **arithmetic logic unit** (ALU). More-complicated mathematical operations (square root, exponentiation, etc.) can be done by repeated operations of the ALU under the control of a program stored in memory.

To make the ALU perform the proper operation at the proper time we need additional circuitry to decode the instructions and to tell the ALU what to do next. We also need a number of storage registers in which intermediate results of calculations performed by the ALU can be stored. Another register, called the **memory address register** (MAR), is required to keep track of what location in memory is being addressed, and a **memory data register** (MDR) would contain what is found or what is to be stored at that address. One very important register is the **program counter** (PC) which keeps track of the memory address of the current program instruction. After each instruction is executed, the program counter is incremented, thereby providing the address of the next instruction to be executed.

An internal clock is also required to regulate the rate at which the various operations are performed. The clock cycle usually has at least two distinct phases. During the **fetch** phase, the instructions are obtained from memory and the appropriate function-select lines to the ALU are energized. During the **execute** phase, the arithmetic operation is performed. The speed of a computer is thus determined by multiplying the cycle time (typically $0.1–1\ \mu s$) by the number of individual arithmetic operations that have to be performed. There is a tradeoff between complexity and speed. A simple computer with a limited instruction set and a limited number of storage registers will have to perform more operations and will take longer to complete a given task than one in which many instructions and registers are available.

An ALU with its associated decoding circuitry, registers, and internal clock is called a **central processing unit** (CPU). The CPU is the heart of the computer. It is a tribute to integrated circuit technology that an entire CPU can be fabricated on a single chip of silicon in less than $1\ cm^2$ of area, and at a cost not very different from the cost of a single vacuum tube. Such an integrated circuit CPU is called a **microprocessor**. Probably no single device since the invention of the transistor has done more to revolutionize the way in which electronics influences our lives.

Finally, the computer needs some means for communicating with the outside world. For this purpose we require **I/O** (input/output) devices. Input to the computer is typically by a keyboard, punched cards, punched tape, magnetic tape, or an A-to-D converter. Output typically involves a teletype, line printer, CRT display, D-to-A converter, or X-Y plotter. These devices are often called **peripherals**. Large computers invariably have various forms of peripheral memory such as magnetic tapes, discs, or drums, which can store huge amounts of information (typically many megabytes) but which require the order of a second rather than the order of a microsecond to access.

The basic components of the computer can be connected together in various ways. The interconnection is called a **bus**, and it usually consists of a multiconductor transmission line, with each conductor carrying one bit of information at a time. The implementation of the bus is a critical decision in the design of a computer. Figure 11.17 shows two possible ways in which the components can be connected. In (a) there is a separate bus to connect the CPU to the memory and to the I/O devices so that everything passes through the CPU. In (b) there is a single bus through which any device can communicate directly with any other. Although the use of a single bus generally allows faster transfer of information, special care must be taken to ensure that only one device at a time is trying to drive the bus. For this purpose **tristate logic** is used in which a third high impedance state is available in addition to the low impedance 0 and 1 states for those circuits that have outputs connected to the bus.

Another important design decision is the choice of serial (bit by bit) or parallel

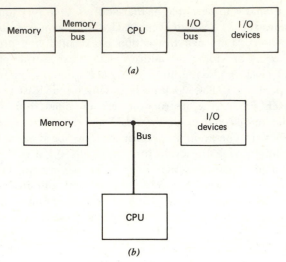

Fig. 11.17 Two possible ways to organize the components of a digital computer. (*a*) Separate buses. (*b*) Single bus.

(byte by byte or word by word) operation of each device. Serial operation reduces the circuit complexity but slows down the execution speed. Usually a compromise is made with inherently slow devices (such as many of the I/O devices) using serial operation and the high-speed devices (such as the ALU) using parallel operation.

The usefulness of a computer can be further enhanced by a number of techniques. One example is the use of **microcoding** in which one instruction invokes a long series of instructions that might be stored in a separate fast memory and executed in its own fast processor so as to produce a result in only a few CPU clock cycles. Another example is the use of **interrupts**, which allow an I/O device to halt the processor's present task and demand that it be serviced. This requires a **priority arbitration** system to decide what can interrupt what and when. Additionally, the processor would have to remember what it was doing so that it could continue after the interrupting task is over.

The I/O devices are connected to the processor through **interfaces**, which resolve voltage and speed differences between the devices as well as initiating interrupts. Although much of the traffic between the I/O devices and the memory would proceed through the CPU, certain devices, by virtue of their high speed, might be allowed direct access to memory. Such interfaces are called **direct memory access** (DMA) interfaces. Interfaces are sometimes provided with limited computing power to relieve the processor of the time-consuming task of controlling complicated I/O devices. These interfaces are generally called **controllers**.

To program a computer to perform a particular task, it is necessary to provide the computer with an array of binary numbers which the CPU can use to set all the circuits in the proper manner. These binary instructions are called **machine language**. Programming in machine language is extremely tedius, since all the instructions look pretty much the same, and it is easy to make errors. For this reason, a simpler mnemonic language called **assembly language** is used, in which the commands have easy-to-remember names like ADD, MOV (move), and MUL (multiply). The computer then generates the machine-language instructions, using a translation program called an **assembler**. Some assemblers even implement error checking by detecting illegal conditions (divide by zero, etc.), and halting execution with an appropriate error message.

Even assembly language is cumbersome for anyone but a professional programmer, and so various higher-level languages have been developed for general use. Common examples are **BASIC** (beginners' all-purpose symbolic instruction code), **PASCAL** (after Blaise Pascal, a French mathematician who built a successful digital calculating machine in 1642), and **FORTRAN** (formula translator). These languages use familiar statements such as ordinary algebra and patterns of logic that closely resemble human thought. The high-level languages are also relatively machine independent, so that a program written for one computer can be easily run on another.

Such languages tend to fall into one of two categories. BASIC is an example of an **interpretive** language, because it stores the program exactly as it was written, When it is run, the program is executed on a line-by-line basis. No attempt is made to restructure the program or predetermine what functions, tables of symbols, or storage will be needed. Interpreters have the advantage that they require a relatively small amount of memory, and since the program is executed exactly as written, the user has a clear idea what is happening. This is a great advantage when the programmer is writing, testing, or editing a program.

FORTRAN is an example of a **compiled** language. The programmer first writes a program and then uses a program called a **compiler**, which translates the program from the original **source code** to an **object code** the machine can understand. In the process, the compiler will construct symbol tables, allocate storage to variables, check for certain types of errors, and perhaps optimize the code to run faster and more efficiently. The statements required to perform a typical algebraic operation in three different languages are shown in table 11.4.

TABLE 11.4   Comparison of Various Computer Languages

FORTRAN	Assembly Language	Machine Language
		0 001 011 111 000 001
	MOV B, R1	0 001 000 000 000 010
		0 110 011 111 000 001
	ADD C, R1	0 001 000 000 000 100
A = (B + C) * D		0 111 000 001 011 111
	MUL D, R1	0 001 000 000 000 110
		0 001 000 001 011 111
	MOV R1, A	0 001 000 000 000 000
	→	→
	Compiler	*Assembler*

A common feature of high-level languages is that a significant amount of the computer's memory has to be dedicated to the translation programs, leaving correspondingly less for the source program and data storage. Consequently, high-level languages are only compatible with large computers. The programs that are written to enable a computer to act intelligently are collectively referred to as **software**, in contrast with the actual electrical circuits and mechanical devices that constitute the **hardware**. In a modern computer, the human effort and expense involved in software development is comparable to that required for the hardware. In the early days of computers, the software was given away in order to sell the hardware. The day may come when the hardware is so inexpensive that it will be given away in order to sell the software, which still requires much human thought and ingenuity.

Recent advances, especially in semiconductor technology, have brought the size and cost of computers down while increasing their speed and flexibility. Computers are an economic anomaly. Their history shows a continuous drop in cost along with a dramatic increase in computational power. This, of course, encourages the use of computers in many new applications.

One important innovation was the development of the **minicomputer**. These machines, while somewhat limited in speed and computational power, are sufficiently inexpensive and simple as to encourage their use as dedicated controllers (in devices such as machine tools and environmental control systems), as programmable calculators, and as laboratory instruments that collect and analyze scientific data. They are used on board aircraft and ships to handle navigation and other duties.

Increasing density in integrated circuit technology has spawned the **micro-computer**, of which the programmable electronic calculator is one example. The heart of the microcomputer is a microprocessor. The microprocessor differs in part from a full-sized computer CPU in word length. Microprocessors typically use 4- to 16-bit words, whereas computers usually have 16- to 64-bit words. Microprocessors can be used as building blocks for larger computers or to perform a specific operation such as controlling the temperature and timing of an oven or controlling the operation of a clock radio. This revolutionary development has expanded computer applications almost without limit.

The advantages of computer control in a product are numerous. Improvements and other changes can often be made by simply changing the program rather than by costly hardware modifications. Greater control sophistication is encouraged by the ease with which it may be implemented by computer.

Microcomputers are now being produced that can use high-level languages and thus compete favorably with minicomputers. The computer industry is developing at a staggering pace. New technological developments such as **bubble memories** and **charged-coupled devices** (CCD) promise vast amounts of storage at almost inconsequential cost. The availability of almost unlimited memory will probably make the oral programming of computers possible within a relatively short time. New semiconductors promise greater speed and density along with lower power dissipation. Digital logic, especially ECL, is getting so fast that a major barrier to speed is the propagation delay along leads. Some ultra-high-speed computers have had to resort to unusual physical layouts to minimize lead length. Some day, the highest-speed computers will have to be the size and shape of baseballs, or perhaps ball bearings. Already, engineers are beginning to use microprocessors as circuit elements in the same way they have been using transistors, op amps, and digital logic integrated circuits. The computer revolution might well be seen by future historians as an event of comparable importance to the Industrial Revolution.

11.8 Summary

Digital circuits are circuits in which the voltages can take on only two rather distinct values. Such circuits are ideally suited for performing binary arithmetic and logical operations. The building blocks of all digital circuits are the **AND** and the **OR** logic gates and their inverse counterparts, the **NAND** and the **NOR** gates. Actually, all logical operations can be performed using only a single type of gate such as the **NAND** gate. The theorems of Boolean algebra indicate how this is done. The behavior of logic circuits can always be analyzed by writing out the truth table.

Applications of logic circuits extend all the way from the half-adder which can't even add two and two, to the digital computer, which is the most powerful and versatile of the digital circuits. Logic gates can be used to make flip-flops, which are useful in multivibrators, counters, and shift-registers. Finally, circuits can be constructed that convert digital information to analog information and vice versa. Digital devices can be extraordinarily complex, such as the digital computer, but fortunately, integrated circuit technology allows such devices to be produced in small packages at a cost that compares favorably with the vastly simpler circuits of only a few years ago.

Problems

11.1   Find the decimal equivalents of the following numbers: 1101011_2, 3742_8, $A6F_{16}$.

11.2   Find the octal equivalents of the following numbers: 101011101_2, 9274_{10}, $E49B_{16}$.

11.3   Find the binary and hexadecimal equivalents of the following decimal numbers: 297_{10}, 3168_{10}.

11.4   Find the binary coded decimal (BCD) equivalents of the following numbers: 101101101_2, 8735_8, 789_{10}.

11.5   Perform the subtraction $537_{10} - 891_{10}$ using two's complement arithmetic and show that the result is correct.

11.6   Using the rules of long division, divide the binary number 101101111_2 by 1100_2 and show that the result agrees with the result of dividing the equivalent decimal numbers.

11.7   Write down a logical expression for the output of the circuit below. Design a circuit using only **NAND** gates that will produce the same output.

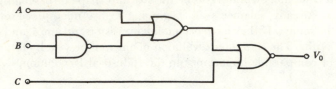

11.8   Use logic gates to design an **anticoincidence circuit**, that is, a circuit that will produce an output whenever a pulse appears at input A, provided that it does not appear simultaneously at input B.

11.9   Write out the truth table for the quantity $A \cdot (\bar{B} + C) + \bar{A}$.

11.10   Use **NAND** gates to perform the following functions:

$$\text{(a)} \quad A + B + \bar{C}$$

$$\text{(b)} \quad A + (B \cdot \bar{C})$$

$$\text{(c)} \quad (A + B) \cdot (\overline{C + D})$$

11.11   Use **NOR** gates to form the functions in problem 11.10.

11.12   Design a monostable multivibrator using **NAND** gates.

11.13   Design an oscillator using **NAND** gates.

11.14   Using the theorems of Boolean algebra, prove the following equalities:

$$A + \bar{A} \cdot B = A + B$$

$$A \cdot B + \bar{A} \cdot B + A \cdot \bar{B} = A + B$$

$$\bar{A} + A \cdot \bar{B} = \bar{A} + \bar{B}$$

11.15   Prove by means of truth tables the equalities in problem 11.14.

11.16   Sometimes a zero voltage is used to represent logical 1 and a positive voltage to represent logical 0. This is called **negative logic**. Design an **OR** and an **AND** gate using resistors and diodes for negative logic.

11.17   Draw the circuit for the RS flip-flop in figure 11.10(a) in terms of individual circuit components (resistors, diodes, and transistors), and convince yourself that it is just an elaborate version of the bistable multivibrator shown in figure 10.6(a).

11.18   Design a binary counter circuit using JK flip-flops that will count from zero to nine and on the tenth count will reset to zero and generate a carry bit. Such a circuit is called a **binary coded decimal** (**BCD**) **counter**, and has obvious applications.

11.19   The device below produces sequences of four-bit binary numbers. If we start at $A = 0$, $B = 1$, $C = 1$, and $D = 0$, find the next five numbers in the sequence which results after each clock pulse.

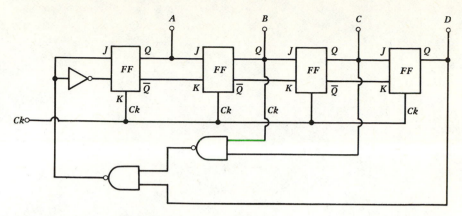

11.20 Construct an *RS* flip-flop using **NAND** gates.

11.21 In the D-to-A converter of figure 11.14(*a*), what accuracy of the resistors is required in order to maintain the inherent accuracy of the digital input? What accuracy would be required for a 16-bit input?

chapter 12

Communications
Electronics

12.1 Electromagnetic Radiation

One of the most interesting predictions of Maxwell's equations of electromagnetism is the existence of **electromagnetic waves** which can propagate through a vacuum with a speed of $c = 3 \times 10^8$ m/s. Electromagnetic waves in free space consist of an oscillating electric and magnetic field oriented at right angles to each other with both perpendicular to the direction of propagation. The frequencies and wavelengths of the electric and magnetic fields are identical, but at a given point in space the fields are 90° out of phase. Electromagnetic waves are unique in that they require no medium for transmission, and yet they transmit energy from one point to another. Electromagnetic waves can occur at any frequency, but for a given frequency f, the wavelength λ in free space is fixed according to the relation

$$\lambda = c/f \tag{12.1}$$

As an electromagnetic wave travels through a medium, such as a piece of glass, its frequency remains constant, but its velocity, and hence its wavelength, decreases. Many forms of electromagnetic waves are distinguished according to how they are generated and detected, but the only fundamental difference between them is their frequency. Figure 12.1 shows the spectrum and the common names and uses of the various frequencies. The reader should note especially the frequencies used for AM broadcasting (535–1705 kHz), FM broadcasting (88–108 MHz), and television (54–890 MHz). The boundaries between the various types of waves are somewhat arbitrary.

One method of producing electromagnetic waves, especially in the radio frequency range (100 kHz–1000 MHz) is with the use of an **antenna**, as shown schematically in figure 12.2. The antenna can be almost anything, such as a simple piece of wire. The antenna in figure 12.2(*a*) is called a **ground plane antenna**, and it typically consists of a vertical element $\frac{1}{4}$ wavelength long rising above the earth or above an artificial conducting plane. The antenna in figure 12.2(*b*) is called a **dipole antenna**, and it typically consists of a $\frac{1}{2}$ wave horizontal conductor fed at its midpoint by a transmission line. For an antenna $\frac{1}{2}$ wave long, the voltages at the ends are 180° out of phase, and it thus resembles an oscillating electric dipole. For the ground plane

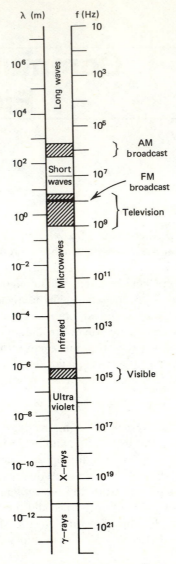

Fig. 12.1 The electromagnetic spectrum.

antenna, the other half of the antenna appears as an image (as in a mirror) below the ground plane. It is thus virtually the same as a horizontal dipole except rotated by 90°. The ground plane antenna radiates the same in all horizontal directions. The dipole radiates best off its broad side (i.e., such that its ends appear to the observer as far apart as possible).

Any antenna will present a complex impedance to an ac voltage source connected to its terminals. A short antenna (short compared with a wavelength) is

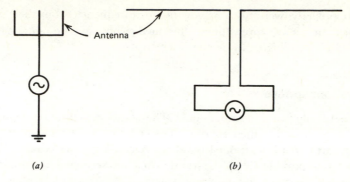

Fig. 12.2 Sinusoidal source connected between: (*a*) an antenna and ground, (*b*) two terminals of an antenna.

mostly capacitive. A long antenna may be either capacitive or inductive, depending on its length, in a manner similar to the open-circuited transmission line (see section 5.6). At certain critical lengths (such as $\frac{1}{4}$ wave), the reactance of the antenna becomes zero, and only a small resistive part remains. This resistance is called the **radiation resistance**. It is typically on the order of $100\ \Omega$. However, unlike an ordinary resistor, it does not represent a dissipation of electrical power in the antenna, but rather a radiation of the energy into the surrounding space. The power is supplied by the source, converted into electromagnetic waves by the antenna, and radiated into space, only to be dissipated by objects some distance away. The dissipation is so slight, however, that a detectable amount of energy usually exists at great distances from a well designed antenna.

In an exactly analogous fashion, an antenna can be used to convert electromagnetic radiation into electrical power which can be delivered to a load connected to the antenna. Such a **receiving** antenna can be thought of as a Thevenin equivalent circuit with a source resistance equal to the radiation resistance. The Thevenin equivalent voltage is a linear superposition of all the nearby sources of electromagnetic radiation. The voltage is usually small (millivolts or less), and so it is important to properly match the radiation resistance to the input resistance of the load for maximum power transfer. Since the antenna is often separated from the transmitter or receiver by some distance, a transmission line is often necessary. An antenna optimized for transmission will also be optimized for reception and vice versa. In fact, the reciprocity theorem demands that under most conditions, for a given pair of antennas, one connected to a transmitter and the other to a receiver, the received signal strength will be the same if the transmitter and receiver are interchanged. For best propagation, the **polarization** (vertical or horizontal) of the two antennas should be the same.

Sometimes antennas of the optimal length are impractically long. A $\frac{1}{4}$-wave antenna for an AM automobile radio would require a length of about 100 m! In such a case, the reactive component of the antenna's impedance can be eliminated by means of some form of resonant *LC* circuit. Such antennas are inherently inefficient and operate properly over only a narrow range of frequencies, but their use is often unavoidable. In many applications, the inefficiency of the antenna is easily compensated for by a high gain in the receiver.

12.2   The Ionosphere

Radio waves usually travel in straight lines. They easily penetrate insulators such as a glass window or a wood-framed building. They tend to be reflected from conductors such as the earth or a steel-framed building. Accordingly, we would expect radio transmission to be possible only if the transmitting and receiving antennas are within line-of-sight, with perhaps only minor obstructions. At very-high-frequencies (above ~ 100 MHz), the propagation of waves is reasonably well described by the above facts. Actually, because of a slight refraction of radio waves by the atmosphere, the radio horizon is usually slightly farther away than the optical horizon.

We know, however, that radio transmissions are possible between distant points on the earth. This occurs because the upper layers of the atmosphere contain a significant density of ionized gas atoms. Such a gas is called a **plasma**, and it is an electrical conductor. The ionization is produced primarily by ultraviolet radiation from the sun. When radio waves travel upward, they can be reflected from this region called the **ionosphere** provided their frequency is below the **plasma frequency**:

$$f_p = 9.0 \sqrt{n} \qquad\qquad (12.2)$$

where n is the number of free electrons per cubic meter.

The electron density in the ionosphere varies with height, time of day, time of year, and sunspot activity. Typically, the peak density occurs at an altitude of several hundred kilometers and has a value of $\sim 10^{12}$ electrons/m^3. Therefore, radio waves are usually reflected if their frequency is somewhat below ~ 10 MHz. At much higher frequencies, the waves usually pass through the ionosphere.

We are now in a position to understand some of the properties of wave propagation. Low-frequency waves such as those in the AM broadcast band (~ 1 MHz) are reflected from the lower layers of the ionosphere, where the density is low. At night, the density drops, and the waves penetrate higher before being reflected, and so the range increases as shown in figure 12.3. Higher frequencies such as the short wave bands (5–25 MHz) have the longest range, depending on the condition of the ionosphere, but too high a frequency, or too low an electron density (such as might occur at night) results in penetration rather than reflection, and then the coverage is limited to line-of-sight, as with FM radio and television. Communication with satellites and deep-space probes must take place at very high frequencies ($\gtrsim 100$ MHz).

The reflected (or **sky**) wave, since it requires special conditions in the ionosphere, tends to provide a much less reliable form of communication than the direct (or **ground**) wave. **Fading** of the signal is common, because ionospheric conditions fluctuate. Intense solar activity can generate a high flux of charged

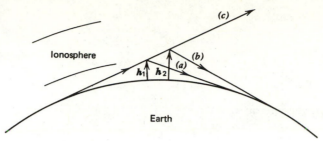

Fig. 12.3 As the frequency is raised or as the ionosphere becomes less dense, a radio wave (*a*) is reflected at a higher altitude (*b*) until finally it is not reflected at all (*c*).

particles that bombard the earth and produce aurora displays, magnetic storms, and ionospheric disturbances. At intermediate distances, the ground wave and sky wave can interfere with each other, causing intense fading as the path length for the sky wave varies.

Multiple reflections of the sky wave are possible, but the energy loss in each such reflection is high. Nevertheless, at very low frequencies ($f \lesssim 10$ kHz), the earth and ionosphere form a waveguide (see section 5.7), and with sufficiently high power, around-the-world propagation can be achieved. These low frequencies are also of interest because the penetration depth into a conducting medium such as saltwater is reasonably great, and they thus provide one of the few reliable means for communicating with underwater submarines.

12.3 Types of Modulation

In the preceding sections we have considered radio waves that consist of a single, sinusoidal frequency component. Such a wave is of limited use because it can carry essentially no information. Generally it is necessary for the sine wave to vary in some manner (such as a variation in its amplitude), usually at an audio frequency (~ 20–20,000 Hz) to transmit speech or music. Such a variation is called **amplitude modulation** (AM), and it is illustrated in figure 12.4(*a*). Note that the audio and radio frequency waves are multiplied together, not added. Thus modulation is an inherently nonlinear operation and requires the use of nonlinear components. To understand how this is done, imagine a two-port network in which the output voltage is given as a function of the input voltage by

$$V_{\text{out}} = aV_{\text{in}} + bV_{\text{in}}^2 \tag{12.3}$$

The first term is the linear term, and the second term represents the nonlinearity. These can be thought of as the first two terms of a Taylor series for an arbitrary nonlinear network. In general, a constant term could be included, but it would be zero for a passive network. Now imagine that two sinusoidal voltages of different frequencies are applied in series at the input:

$$V_{\text{in}} = V_1 \sin \omega_1 t + V_2 \sin \omega_2 t$$

Then the output would be

$$V_{out} = aV_1 \sin \omega_1 t + aV_2 \sin \omega_2 t$$
$$+ bV_1^2 \sin^2 \omega_1 t + bV_2^2 \sin^2 \omega_2 t \qquad (12.4)$$
$$+ 2b^2 V_1 V_2 \sin \omega_1 t \sin \omega_2 t$$

By using the trigonometric identity,

$$\sin^2 \theta = \tfrac{1}{2}(1 - \cos 2\theta)$$

and, rearranging terms, the above equation can be written as

$$V_{out} = \tfrac{1}{2}b(V_1^2 + V_2^2) + aV_1 \sin \omega_1 t$$
$$- \tfrac{1}{2}bV_1^2 \cos 2\omega_1 t - \tfrac{1}{2}bV_2^2 \cos 2\omega_2 t \qquad (12.5)$$
$$+ (a + 2b^2 V_1 \sin \omega_1 t) V_2 \sin \omega_2 t$$

The output then consists of five terms: (1) a dc term, like the output of a rectifier, (2) a sinusoidal term at ω_1, (3) a sinusoidal term at $2\omega_1$, (4) a sinusoidal term at $2\omega_2$, and (5) a sinusoidal term at ω_2 whose amplitude is varied (modulated) by ω_1. If $\omega_1 \ll \omega_2$, as is usually the case for amplitude modulation, it is relatively easy to filter out all but the last term, thereby producing an amplitude-modulated wave. The **percentage modulation** is determined by the ratio $2b^2 V_1/a$, such that when $2b^2 V_1/a = 1$, the wave is said to be 100% modulated, that is, the amplitude just goes to zero at the peak of the modulation. The case shown in figure 12.4(a) is ~50% modulated. A modulation of 100% is usually considered optimal. A lesser amount is wasteful of power, and a greater amount results in severe distortion of the waveform.

The terms in equation 12.4 can be rearranged in a slightly different way by using the trigonometric identity

$$\sin \theta \sin \phi = \tfrac{1}{2}[\cos(\theta - \phi) - \cos(\theta + \phi)]$$

to get the result

$$V_{out} = \tfrac{1}{2}b(V_1^2 + V_2^2) + aV_1 \sin \omega_1 t + aV_2 \sin \omega_2 t$$
$$- \tfrac{1}{2}bV_1^2 \cos 2\omega_1 t - \tfrac{1}{2}bV_2^2 \cos 2\omega_2 t \qquad (12.6)$$
$$+ b^2 V_1 V_2 \cos(\omega_2 - \omega_1)t - b^2 V_1 V_2 \cos(\omega_2 + \omega_1)t$$

As before, Fourier components appear at dc, ω_1, ω_2, $2\omega_1$, and $2\omega_2$, but new components also appear at $\omega_2 \pm \omega_1$. The presence of the second harmonic of ω_1 and ω_2 results from the initial assumption of equation 12.3. A term proportional to V_{in}^3 would have given rise to a third harmonic, and so on. The fundamental frequencies ω_1 and ω_2 can be eliminated by using an appropriately balanced circuit, such as a bridge rectifier for which $a = 0$. For an AM signal, the higher frequency ω_2 is called the **carrier**, and the two frequencies $\omega_2 \pm \omega_1$ are called the **sidebands**. In a typical case the carrier might be at 1 MHz while the modulation might be at 1 kHz. Then

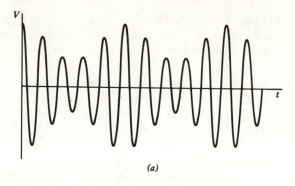

(a)

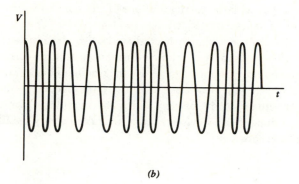

(b)

Fig. 12.4 *(a)* Amplitude modulation. *(b)* Frequency modulation.

the Fourier spectrum of the AM signal consists of a 1000-kHz carrier with sidebands at 999 kHz and 1001 kHz. A radio wave modulated by an audio signal with Fourier components up to, say, ~10 kHz would thus be spread out in frequency in a band ~20 kHz wide centered around the carrier frequency.

An alternate form of modulation, called **frequency modulation** (FM), is illustrated in figure 12.4(*b*). In this case, the frequency of the carrier varies in response to the applied audio signal. The amount by which the frequency varies is called the **deviation**. If the deviation is small, the Fourier spectrum contains only components near the carrier frequency. The exact calculation is complicated, since it depends on both the frequency spectrum of the audio signal and the deviation. The spectrum is usually slightly broader than the corresponding AM signal. The advantages of FM over AM are that nonlinearities in the radio frequency amplifiers in the transmitter and receiver do not affect the quality of the signal, and the FM receiver is relatively free from electromagnetic interference (noise) which tends to be amplitude modulated. The high **fidelity** of an FM broadcast signal comes about not as an inherent quality of FM but rather as a result of the use of larger bandwidths that allow the transmission of much higher frequency Fourier components of the audio signal. An AM signal with equal bandwidth would be equally clear and crisp.

With either type of modulation, the circuits in the transmitter and receiver must be capable of passing a band of frequencies centered on f_0 and wide enough to accommodate the sidebands without attenuation. On the other hand, a narrow bandwidth is desired to avoid interference from strong stations on adjacent frequencies, as well as to permit the reception of weak signals in the presence of noise (see section 9.7). A radio receiver always represents a compromise between these conflicting requirements.

Audio signals are not the only kind of information that can be transmitted on a carrier wave. A television video signal is transmitted in the same way (see section 12.6), as are various forms of digital data. Other more exotic forms of modulation are also in wide use. For example, **single sideband** (SSB) modulation, which is basically AM with the carrier and one of the sidebands removed, allows the same information to be transmitted in a bandwidth only half as wide as an AM signal and with more efficient use of the power radiated by the transmitter. This is possible because the two sidebands carry the same information, and the carrier contains most of the power but no information. To generate the original Fourier components of the audio signal, however, it is necessary to reinsert the carrier at the receiver by a nonlinear process similar to the one by which the modulated wave was originally generated. This is done with a device called a **beat frequency oscillator** (BFO) or with a **product detector**.

12.4 Radio Transmitters

A radio transmitter consists of an oscillator, one or more amplifier stages, and a modulator as shown in figure 12.5. The oscillator is usually crystal-controlled to improve its frequency stability. Frequency multiplier stages are often placed between the oscillator and amplifier. The amplifier is normally operated class C to improve efficiency, since only a narrow band of frequencies must be amplified. The **modulator** can be simply a switch to turn the carrier on and off for transmitting Morse code.

In an AM transmitter, as shown in figure 12.5(a), the modulation is usually applied at the final amplifier stage to eliminate distortion caused by the nonlinearities of the class C amplifier. The modulator usually contains a class A or push-pull class B audio amplifier which controls the operating point of the radio frequency amplifier. It must provide an amount of audio power equal to about half of the radio frequency carrier wave. Sometimes amplitude modulation is applied to a stage prior to the final amplifier to reduce the amount of power that must be supplied by the modulator. In such a case, the subsequent amplifier stages must be operated in a linear fashion, thereby reducing their efficiency.

In the FM transmitter shown in figure 12.5(b), the modulator varies the oscillator frequency, for example by means of a varicap diode in the oscillator LC circuit. Such a modulator need supply very little power, and the subsequent radio

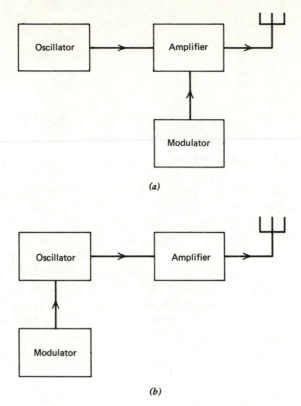

Fig. 12.5 Transmitters. (*a*) AM. (*b*) FM.

frequency amplifier stages need not be linear, but they must have adequate bandwidth to accommodate the Fourier spectrum of the oscillator signal.

12.5   Radio Receivers

The function of a radio receiver is to extract the audio signal from a modulated radio frequency carrier wave. The simplest radio receiver would consist of nothing more than an antenna, a diode (preferably germanium), and a pair of headphones. Such a receiver possesses little or no frequency **selectivity**, and will usually receive several of the strongest AM broadcast stations simultaneously. By simply adding a variable inductor that forms a resonant LC circuit with the antenna capacitance, as shown in figure 12.6, a quite usable radio receiver can be constructed which is capable of tuning in a number of different AM stations.

Like modulation, the process of **demodulation** or **detection** is inherently nonlinear, thereby accounting for the necessity of the diode in figure 12.6. Recall that all of the Fourier components of a modulated radio wave lie in a small band of frequencies near the frequency of the carrier wave. The diode generates the low-

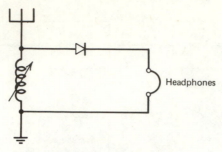

Fig. 12.6 Simple diode receiver.

frequency (audio) Fourier components which are absent in the transmitted wave. The operation is similar to a diode rectifier in which a low-frequency (dc) component is generated out of a higher-frequency (60-Hz) ac component. Figure 12.7 shows how this is done. Figure 12.7(a) shows an amplitude-modulated wave. After passing through the diode, only the positive half of the wave remains, as shown in figure 12.7(b). This wave clearly contains low-frequency components in addition to components at the frequency of the carrier wave. With the addition of a low-pass filter, only the low-frequency components remain, as shown in figure 12.7(c). A dc component is also present, but this is easily eliminated. For the case of the simple radio receiver in figure 12.6, the frequency response of the headphones and of the human ear provide the low-pass filter. The ear is actually an extraordinarily good filter, attenuating frequencies below ∼20 Hz and above ∼20 kHz by an amount that would be difficult to approach by the use of electrical components.

The simple diode receiver can be improved by adding radio frequency amplifiers between the antenna and the diode so that a smaller antenna can be used, additional frequency selectivity provided, and the nonideal properties of the diode overcome. Similarly, audio frequency (AF) amplifiers can be used after the diode so that the output can drive a **loudspeaker**.

Although the scheme described above will work, it is seldom used. To produce adequate frequency selectivity, many resonant circuits are required, and it is difficult to keep them all tuned to precisely the same frequency as the receiver frequency is varied. A more common scheme is to use a **superheterodyne circuit** as shown in figure 12.8. The incoming radio frequency signal (1000 kHz in this example) is used to amplitude modulate a signal from a variable frequency **local oscillator** (1455 kHz in this case) in a circuit called a **mixer** or **converter**. It produces an output equal to the sum and difference of the two frequencies (455 kHz and 2455 kHz), as was shown in equation 12.6. These two frequencies can be thought of as sidebands of the 1455-kHz amplitude-modulated oscillator, although they are quite widely separated. The 455-kHz component is amplified by a highly selective, fixed frequency, **intermediate frequency** (**IF**) **amplifier**, and then fed into a circuit called a detector, which extracts the modulating signal in the same way as the diode in figure 12.6. The 2455-kHz signal is far outside of the passband of the IF amplifier and thus is rejected.

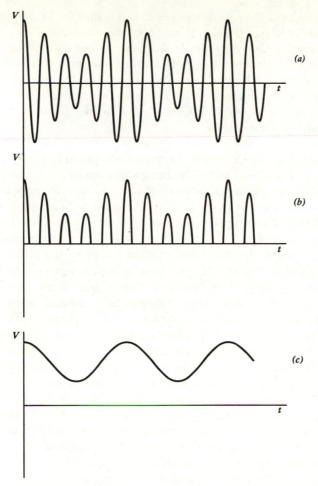

Fig. 12.7 The amplitude modulated wave in (a) contains
no low frequency components, but such components can be
generated by use of a diode (b) and subsequent low-pass
filter (c).

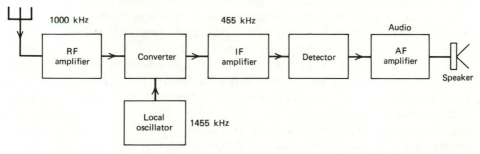

Fig. 12.8 Superhetrodyne receiver.

The advantage of the superhetrodyne circuit is that the IF amplifier frequency can remain fixed, and thus it is easy to maintain a narrow bandwidth for a wide range of input frequencies. Also, the lower IF frequency means that for a given circuit Q, the bandwidth is smaller, since $\Delta f = f_0/Q$. Furthermore, since f_0 is independent of the frequency of the received signal, a constant selectivity can be maintained over an unlimited range of received frequencies. Lower frequency amplifiers are also easier to construct so as to avoid unwanted oscillation and the like, because stray inductances and capacitances are less critical.

A superhetrodyne receiver is not without its difficulties, however. For example, note that a signal at 1910 kHz will also produce a 455-kHz IF signal for the case shown in figure 12.8. This is called the **image frequency**, and it is always present, separated from the frequency of the received signal by an amount equal to twice the IF frequency. Thus a strong signal at 1910 kHz might be heard when the receiver is tuned to 1000 kHz. The image is suppressed by providing adequate selectivity in the radio frequency amplifier stages preceding the converter. As the radio frequency becomes higher, image rejection becomes more and more of a problem. For a 100-MHz receiver with a 455-kHz IF, the signal and its image are separated by only $\sim 1\%$, and adequate selectivity is difficult to obtain while still allowing the receiver to tune over a wide frequency range. Consequently, **double conversion** super-hetrodynes are sometimes used, in which the first IF (perhaps 5 MHz) provides the image rejection and the second IF (often 455 kHz) provides the selectivity.

The presence of an oscillator in a superhetrodyne receiver poses certain other difficulties. Care must be taken to isolate the oscillator from the receiving antenna, so that the receiver does not act as a transmitter and generate unwanted interference. It is also possible for harmonics of the local oscillator to interfere with the incoming signal at certain frequencies if the oscillator is on the low side of the signal frequency. In a dual-conversion receiver with two oscillators, a great many spurious frequencies can be generated by the beating together of various harmonics of the oscillators. A certain care must therefore be exercised in choosing appropriate IF frequencies, in maintaining good sine waves (low harmonic content), and in isolating the oscillators from one another.

Another consideration in the construction of any AM receiver is the fact that a wide variation in signal strength usually exists for the various stations to which the receiver might be tuned. This fact would only be an annoyance, causing a variation in the audio volume as the receiver is tuned from one station to another, except that it is difficult to design amplifiers that have a sufficiently large amplification for weak signals without causing distortion to the strong signals because of nonlinearities in the amplifiers. Consequently, most AM receivers have an **automatic volume control** (AVC), more properly called an **automatic gain control** (AGC), whose function is to reduce the gain of the radio and intermediate frequency amplifiers by an amount that increases with the strength of the received signal. This is usually done by taking the dc component of the detector output [see figure 12.7(c)], and using it to control the bias and hence the operating point of the preceding amplifiers. An AGC is thus a form of negative feedback, but at very low frequency.

The AGC voltage also provides a convenient means for monitoring the strength of the received signal. Such an **S-meter** is labeled in **S-units** (S1 to S9) and dB over S9 and is usually calibrated so that a voltage of 50 μV at the receiver input gives a reading of S9, and each S unit corresponds to 6 dB. S-meters are rarely calibrated with great accuracy, but they do provide an extremely useful indication of relative signal strength. An AM detector circuit containing volume control, AGC, and S-meter functions is shown in figure 12.9.

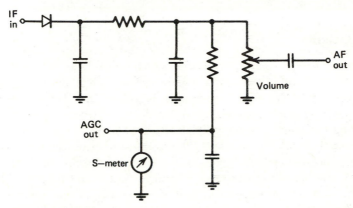

Fig. 12.9 AM detector circuit, containing volume control, AGC. and S-meter.

An FM receiver is also usually a superheterodyne, but its detector contains a frequency-selective LC filter that produces an output voltage proportional to the deviation of the signal frequency from its unmodulated value. Since the audio volume of an FM signal is independent of the signal strength and depends only on the frequency deviation, the use of an AGC circuit is unnecessary. In fact, FM receivers are normally designed with such high amplification that the last IF stage is driven from cutoff to saturation for even a very weak signal. Such a saturated amplifier is called a **limiter**, and it serves to make the output quite insensitive to amplitude variations. Because of the large amplification, an annoying amount of noise is usually present in an FM receiver in the absence of a signal. A **squelch** circuit is thus often provided to suppress the audio output noise in the absence of an input signal.

Because of the high frequencies used for FM broadcasting ($\sim$100 MHz), even a small percentage drift in the local oscillator frequency can be quite objectionable. Consequently, most tunable FM receivers have an **automatic frequency control** (AFC) which detects the variation of the IF signal from the center of the IF passband and produces a corresponding dc voltage which is used to stabilize the local oscillator frequency. An AFC circuit is yet another example of negative feedback.

Whereas the bandwidth of a commercial AM broadcast signal is about 6 kHz, the bandwidth of an FM broadcast signal is about 36 kHz. The wide bandwidth not only improves the fidelity of an FM signal but also permits the transmission of **stereophonic** information by means of a modulated **subcarrier** displaced 19 kHz from the main carrier. Although stereo could be achieved by modulating each carrier with one of the two audio channels, this would cause a monophonic receiver to receive only a single channel. The two systems are made compatible by modulating the main carrier with a signal proportional to the sum of the two channels and modulating the subcarrier with a signal proportional to the difference between the left and right channels. A stereo receiver then reconstructs the two channels by taking the appropriate sums and differences of the two signals.

12.6 Television

The **video** (picture) portion of a television signal is generated by a television (TV) camera which is the optical analog of the microphone. The type of TV camera used for live commercial television is the **image orthocon** in which the picture to be transmitted is imaged on a **photocathode** which emits electrons proportional to the amount of light striking each area of the cathode. These electrons strike a glass disc, causing it to emit secondary electrons. The secondary electrons are collected on a fine mesh screen, leaving the disc with a positive charge density proportional to the light intensity. A low-velocity electron beam from a cathode ray tube is then directed toward the positively charged disc, and the intensity of the reflected beam is measured by a photomultiplier tube (section 10.6). Some of the electrons are extracted from the beam to neutralize the positive charge, and so the detected signal is modulated by the light intensity.

A second type of TV camera which is becoming increasingly popular because of its small size and simplicity is the **vidicon**. It consists of a thin layer of photoconductive material deposited on a transparent conducting film on which the image of the picture is focused. The surface is initially uniformly charged, but those areas on which the light falls discharge at a more rapid rate. Then, when a low-velocity electron beam is swept across the surface, a charging current flows to the conducting film in proportion to the light intensity.

For either type of camera, a well-focused electron beam sweeps across the target in a pattern as shown in exaggerated form in figure 12.10. The beam sweeps slowly from left to right and then is turned off (**blanked**) during the rapid **retrace** when the beam moves back to the left. Actually, rather than sweeping through all 525 lines in sequence, the odd-numbered lines are first swept, and then the trace returns to the top and sweeps through the even-numbered lines. This **interlacing** reduces the bandwidth required for transmitting a picture without objectionable flicker. Even so, it is necessary to transmit 30 complete pictures per second, so the horizontal sweep frequency is $525 \times 30 = 15{,}750$ Hz. To obtain reasonable horizontal resolution, the video bandwidth must be several hundred times greater than this, so that the total bandwidth is about 4.5 MHz. Contrast this with the fact that the whole AM broadcast band is only about 1 MHz wide!

A television transmitter is essentially the same as a radio transmitter, except that the carrier is amplitude modulated with a 4.5-MHz-wide video signal, and an FM

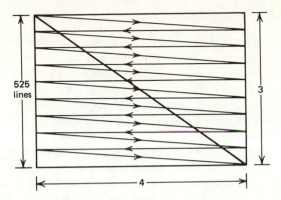

Fig. 12.10 Scan sequence of a television picture.

audio subcarrier is added 4.5 MHz above the video carrier. To conserve bandwidth, the Fourier components above about 1 MHz are suppressed from the lower sideband Since only a vestige of the sideband remains, the process is referred to as **vestigal sideband modulation**.

A television receiver is a superheterodyne with an FM audio section and an AM video section. The detected video signal is fed to a picture tube, which is essentially an intensity-modulated CRT, usually with magnetic rather than electrostatic deflection. The sweep of the picture tube must, of course, be synchronized with the sweep of the TV camera tube. This is done by special horizontal and vertical **sync pulses** which are transmitted during the retrace of the beam.

With color television, three separate camera tubes are used with red, green, and blue filters. To make the system compatible with black-and-white television, a signal, called the **luminance signal**, is produced by adding the outputs of the three color signals. This signal then amplitude-modulates the main video carrier. Then, in a manner similar to the way in which an FM stereo signal is transmitted (see section 12.5), two **chrominance** signals are generated by taking a linear combination of the differences of the three color signals. To keep the bandwidth of the color TV signal comparable to that of black-and-white, use is made of the fact that most of the Fourier components of the luminance signal occur near harmonics of the horizontal sweep frequency, 15,750 Hz. The chrominance signals are transmitted on two subcarriers $90°$ out of phase and having a frequency 3.579545 MHz above the picture carrier so that their harmonics fall midway between the harmonics of the luminance signal. In the usual color television receiver, the red, green, and blue signals are reconstructed and used to control three electron guns which are directed at the viewing screen through a **shadow mask** consisting of a metal sheet with thousands of accurately located holes. The mask allows each beam to illuminate a phosphor dot of the appropriate color on the screen. When viewed from a distance, the dots are indistinguishable and produce a shade of color that depends on the relative intensities of the various dots.

12.7 Radar

An interesting application of electromagnetic radiation is **radar** (*radio detection and ranging*), in which radio waves in the microwave range are reflected off a distant object and detected in order to determine the position and/or velocity of the object. Microwave frequencies ($10^9 - 10^{12}$ Hz) are required, because they can be confined to narrow beams by modest-sized antennas and because they travel in straight lines and penetrate clouds and other nonconducting obstructions.

There are two basic types of radar. With **pulsed radar**, a short (typically 1 μs) pulse of radiation is transmitted at a rate of typically 1000 pulses/s by a rotating antenna. The position of an object is determined from the direction of the antenna and the time required for the pulse to be reflected from the target and return. The position is displayed on a device called a **plan position indicator** (PPI) which is just a cathode-ray tube with a circular sweep that is in sync with the rotation of the antenna. Such a radar is two-dimensional, since it provides bearing and range information only. Its most common application is in air traffic control.

The other type of radar is called **doppler radar**. It makes use of the fact that an electromagnetic wave is shifted in frequency when it is reflected from a moving object. The frequency shift is given by

$$\Delta f \simeq \frac{2fv}{c} \tag{12.7}$$

where v is the component of velocity of the target *toward* the radar and c is the velocity of light ($v \ll c$). The radar receiver mixes the transmitted signal and the received signal to produce a **beat** frequency equal to Δf. The frequency difference is then converted to a voltage that can be read with a dc voltmeter. The doppler radar need not be pulsed and is typically run continuously. Its most common application is in speed detection for law enforcement.

The two types of radar can be combined to discriminate against stationary targets so as to reduce clutter on the radar screen caused by trees, buildings, and the like. Higher-frequency radars ($\lambda < 1$ cm) give higher spatial resolution, but they tend to be strongly attenuated, especially in areas of high precipitation. Weather radars intentionally use frequencies at which significant reflection occurs from regions of precipitation. The range of a typical radar is limited to a few hundred miles, but radio waves have been reflected from the moon and even from several of the closer planets.

12.8 Summary

When applied to an antenna, high-frequency oscillating voltages will generate electromagnetic waves that propagate over large distances without wires or other medium. These waves are characterized by their frequency or corresponding wavelength. They provide a highly effective means of communicating speech, pictures, or other types of information. At low frequencies ($\lesssim 10$ MHz), radio waves reflect from the ionosphere and can travel around the earth. At high frequencies

($\gtrsim 10$ MHz), they usually penetrate the ionosphere and travel deep into space, gradually losing their energy density.

A radio transmitter consists of an oscillator, an amplifier, and a modulator that varies either the amplitude or frequency of the wave. A radio receiver can be nothing more than a diode and a pair of headphones, but more typically it contains a number of amplifiers to improve its sensitivity, and, in the case of the common superheterodyne receiver, one or more frequency converters.

The processes of rectification, amplitude modulation, detection, hetrodyning (beating), and frequency multiplication are all nonlinear operations that generally result from passing one or more signals through any nonlinear circuit. Whether a circuit is called a rectifier, a modulator, a detector, a mixer, or a multiplier depends on which of the operations the circuit is optimized for. By choosing appropriate nonlinear devices and connecting them in a suitable way with filters, any one of the operations can be emphasized and the other suppressed.

Two important examples of the use of electromagnetic waves are television and radar. A television picture is produced by scanning an electron beam in a prescribed sequence and amplitude modulating a carrier in proportion to the light intensity. Color television uses three such beams to record the intensities of each of the primary colors: red, green, and blue. A radar transmits a wave and detects either the time lapse (pulse radar) or the frequency shift (doppler radar) of the wave reflected from a target. In this way the position and velocity of the target can be determined.

Problems

12.1   Calculate the wavelengths corresponding to the following frequencies: (a) an AM radio station at 1070 kHz, (b) an FM radio station at 88.7 MHz, (c) a UHF television station at 700 MHz, (d) a citizen's band transmitter at 27.185 MHz, (e) a microwave oven at 2450 MHz.

12.2   Use dimensional arguments to show that in free space ($Q = 0$, $I = 0$), Maxwell's equations require

$$\frac{E}{B} = \frac{1}{\sqrt{\mu_0 \varepsilon_0}} = c$$

12.3   You notice that the antenna on a police car is about 18 in. long. Assuming it to be $\frac{1}{4}$ wavelength, estimate the frequency of the police radio.

12.4   Assume an antenna has a capacitive reactance corresponding to 1000 pF. Design a circuit that will cause the antenna to look purely resistive to a source at a frequency of 1 MHz.

12.5   A television tower is 1000 ft high. Estimate the range of its coverage if the receiving antenna is at ground level. How high must the receiving antenna be to double the range?

12.6   Estimate the range that could be expected for a radio wave that reflects from the ionosphere at a height of 100 km.

12.7   Calculate the amplitude and frequency of the Fourier components of the amplitude modulated wave given by $V(t) = V_0 \cos \omega_0 t (1 + \cos \Delta \omega t)$.

12.8   Calculate the ratio of the power in the sidebands to the power in the carrier for a 100% modulated AM signal. Assume power is proportional to the square of the voltage.

12.9   Write a mathematical expression for the time dependence of a sinusoidal voltage of peak value 10 V and 100 MHz which is frequency modulated by a 1-kHz sine wave with a deviation of ± 10 kHz.

12.10   FM broadcast receivers often have a 10.7-MHz IF. If such a receiver is tuned to 88.1 MHz, what are the two frequencies for which images might be expected? What determines which of the two frequencies will, in fact, be present?

12.11   Assume an AM broadcast receiver (535–1705 kHz) has a 455-kHz IF and a local oscillator on the low side of the received signal (very unusual). Over what range of frequencies must the oscillator tune? At what frequencies on the radio dial might one expect to hear harmonics of the oscillator? At what frequency on the radio dial might the fundamental of the local oscillator be picked up directly by the IF amplifier? Give three reasons why the local oscillator is usually at a frequency above the received frequency in such a receiver.

12.12   A double conversion superhetrodyne has a first IF of 10.7 MHz and a second IF of 455 kHz. If the receiver is tuned to 121.5 MHz, what are the possible frequencies of the first and second local oscillators?

12.13   A "ghost image" is often observed on a television screen slightly displaced horizontally from the main picture. Usually this is caused by a reflected wave which arrives at the TV antenna slightly later than the main wave. Calculate the difference in path length of the direct and reflected wave if the ghost is displaced by one inch on a screen with a width of 16 in.

12.14   Disregarding the sync pulses, calculate the frequencies at which most of the Fourier components of the TV video signal occur for a carrier frequency of 61.25 MHz (Channel 3) if the picture consists of (a) a single vertical stripe, and (b) a single horizontal stripe.

12.15   Consider a pulsed radar which operates at a frequency of 10 GHz and transmits 1-μs-wide pulses at a rate of 1000 pulses/s. Calculate (a) the number of waves in a pulse, (b) the spatial length of a wavetrain, (c) the spatial distance between pulses, and (d) the minimum bandwidth of the radar signal.

12.16   What is the maximum range of a pulsed radar that transmits 2000 pulses per second? If the pulse length is 0.5 μs, what is the approximate range resolution?

12.17   What frequency shift would be produced in a doppler radar by a target moving at 55 mi/hr if the radar uses 1-cm microwaves?

Appendixes

Unfortunately, all the theory, data, and practices that are needed to solve electronic problems and design circuits have not been compiled in one reference source. This book provides but a starting point for circuit design and analysis. The serious student of electronics will require much more detailed treatments of the various topics. There are hundreds of good sources, but only a handful of the most useful are listed here, along with brief comments on their contents. In addition to these sources, the reader should be aware of the volumes of specifications and application data available free or at nominal cost from electronics manufacturers, especially the semiconductor manufacturers. The electronics magazines (*Electronics, Electronic Design News, Electronic Design, Electronic Products,* and *Digital Design,* to name a few) are very helpful in keeping up with new products, "the state of the art," and engineering practice.

Study other's designs and methods, but don't blindly copy what you don't fully understand. Many published designs either contain errors or have never been fully tested. Perhaps because of the fundamental perversity of Nature, even the most subtle and sophisticated design methods often yield circuits requiring substantial amounts of "bench engineering" to obtain satisfactory performance.

Halliday, David, and Resnick, Robert, *Physics*, Wiley (1978). The starting point for any study of electronics is an understanding of the underlying physical principles. This standard introductory physics text contains a complete, yet not overly involved, explanation of all the important physical processes that govern the behavior of electronic components. A book of this sort should be in the library of every scientist and engineer.

Thomas, George B. Jr., and Finney, Ross L., *Calculus and Analytic Geometry*, Addison-Wesley (1979). The difference between someone who merely tinkers with electronics and someone who can successfully design complicated circuits lies largely in the person's level of mathematical ability. This freshman calculus text provides all the mathematical tools that are necessary for design and analysis of even the most complicated electrical circuits. Besides differentiation and integration, the sections on determinants and linear equations, Fourier series, complex numbers, and differential equations are especially useful in the study of electronics.

Brophy, James J., *Basic Electronics for Scientists*, McGraw-Hill (1977). This standard electronics text contains a good description, often simplified, of most of the topics covered in this book. For the reader who would like to see the topics treated here explained in different (and perhaps better) words, Brophy provides a highly readable text.

Anderson, L. W., and Beeman, W. W., *Electric Circuits and Modern Electronics*, Holt, Rinehart and Winston (1973). This text, which is presently out of print but available in libraries, is probably closer in level, order of topics, emphasis, and notation to the present text than any other electronics book available. It was for several years used as the text in the course for which the present book was written. It delves in much greater detail into many of the topics presented here.

Diefenderfer, A. James, *Principles of Electronic Instrumentation*, Saunders (1979). This relatively modern text, as the title suggests, is slanted toward the use of analog and digital circuits in electronic instruments. It contains a large number of sample specification sheets for semiconductor devices and integrated circuits.

Langford-Smith, F., ed., *Radiotron Designer's Handbook*, distributed by Radio Corporation of America (1965). Although many people will scoff at this relic of the tube days, it is filled with valuable information on almost every electronic subject except semiconductors. Every chapter is followed by a thorough bibliography with detailed information on many subjects.

Westman, H. P., ed., *Reference Data for Radio Engineers*, Sams & Co. (1975). In many ways the modern version of the *Radiotron Designer's Handbook*, this volume contains most of the reference information any electronics designer needs. Though it is slightly tilted toward the communications engineer, most of the articles are of general interest. It is excellent for reference and review but not a very good source for learning new material. It contains a great wealth of tables and formulas.

Tremaine, Howard M., *Audio Cyclopedia*, Sams & Co. (1969). This is a very comprehensive source of information on audio recording and reproduction. Unfortunately, it is organized in a question-and-answer format, which can be annoying, but it has a fairly good index. It is a good source of audio engineering practice, although it contains nothing on semiconductors.

The Radio Amateur's Handbook, American Radio Relay League, published yearly. The publications of the ARRL can be very useful, even to the non-ham, in providing basic, clearly written instruction in electronic principles and good construction practices. *The Radio Amateur's Handbook* is strongly recommended for the beginner in electronics, especially for someone with only a modest mathematical training. It is also relatively inexpensive.

Graeme, Jerald G., Tobey, Gene E., and Huelsman, Lawrence P., eds., *Operational Amplifiers: Design and Application*, McGraw-Hill (1971). There are many excellent textbooks on op amps as well as a wealth of application data available from manufacturers. This book, however, along with its companion volumes by Graeme, Jerald G., *Applications of Operational Amplifiers: Third Generation Techniques*, by Wong, Yu Jen, and Ott, William E., *Function Circuits: Design and Applications*, and Graeme, Jerald G., *Designing with Operational Amplifiers: Applications Alternatives*, provide the best reference on op amp technique available. Written under the aegis of Burr-Brown, a major supplier of operational amplifiers, this reference provides clear, complete information on almost every aspect of op amp application.

Taub, Herbert, and Schilling, Donald, *Digital Integrated Electronics*, McGraw-Hill (1977). For one who was intrigued but frustrated by the brief treatment of digital electronics in Chapter 11, this book expands to over 600 pages on the fundamentals of digital circuits, exclusive of computers. Written in textbook style, with many examples and problems, this book begins where the present text leaves off in digital circuits.

Cannon, Don L., and Lucke, Gerald, *Understanding Microprocessors*, Texas Instruments Learning Center (1979). Although texts on digital computers and microprocessors have a way of becoming outdated by the time they make it to press, this elementary text is as good a starting point as any for one who wants to understand in more detail the construction and operation of microcomputers.

Scientific American, September 1977 issue. This entire issue was devoted to the subject of microelectronics and contains eleven articles on topics ranging from the construction of integrated circuits to the organization and use of microcomputers. It also contains an extensive bibliography.

B
Physical Constants

Gravitational constant:	$G = 6.672 \times 10^{-11} \text{ N} \cdot \text{m}^2/\text{kg}^2$
Acceleration due to gravity:	$g = 9.81 \text{ m/s}^2$
Mean radius of the earth:	$R_\oplus = 6371 \text{ km}$
Mass of an electron:	$m_e = 9.10953 \times 10^{-31} \text{ kg}$
Electronic charge:	$e = 1.60219 \times 10^{-19} \text{ C}$
Velocity of light:	$c = 2.99792 \times 10^8 \text{ m/s}$
Boltzmann's constant:	$1.38066 \times 10^{-23} \text{ J/K}$
Permeability of free space:	$\mu_0 = 4\pi \times 10^{-7} \text{ N/A}^2$
Permittivity of free space:	$\varepsilon_0 = 8.85419 \times 10^{-12} \text{ C}^2/\text{N} \cdot \text{m}^2$
Planck's constant:	$h = 6.62618 \times 10^{-34} \text{ J/Hz}$

C
Units and Conversion Factors

Some metric prefixes: $p = 10^{-12}$ (pico)
$n = 10^{-9}$ (nano)
$\mu = 10^{-6}$ (micro)
$m = 10^{-3}$ (milli)
$c = 10^{-2}$ (centi)
$k = 10^{3}$ (kilo)
$M = 10^{6}$ (mega)
$G = 10^{9}$ (giga)

1 ampere (A) = 1 C/s
1 volt (V) = 1 J/C
1 ohm (Ω) = 1 V/A
1 siemens ($\mho$) = 1 A/V
1 watt (W) = 1 J/s
1 farad (F) = 1 C/V
1 henry (H) = 1 Wb/A
1 hertz (Hz) = $1/2\pi$ rad/s
π radians = 180°
Temperature (Kelvin) = $T(°C) + 273.16$
1 micron (μ) = 10^{-6} m
1 inch = 2.54 cm
1 mile = 1.60934 km
1 calorie = 4.184 J
1 electron volt (eV) = 1.6×10^{-19} J
1 horsepower = 745.7 W

Electromagnetism

Maxwell's equations:

$$\oint \mathbf{E} \cdot \mathbf{dA} = Q/\varepsilon_0 \qquad \text{(Gauss's law)}$$

$$\oint \mathbf{B} \cdot \mathbf{dA} = 0$$

$$\oint \mathbf{B} \cdot \mathbf{dl} = \mu_0 I + \mu_0 \varepsilon_0 \frac{d}{dt} \int \mathbf{E} \cdot \mathbf{dA} \qquad \text{(Ampere's law)}$$

$$\oint \mathbf{E} \cdot \mathbf{dl} = - \frac{d}{dt} \int \mathbf{B} \cdot \mathbf{dA} \qquad \text{(Faraday's law)}$$

Current:

$$I = \frac{dQ}{dt}$$

Voltage:

$$V = \int \mathbf{E} \cdot \mathbf{dl}$$

Magnetic flux: $\Phi = \int \mathbf{B} \cdot \mathbf{dA}$

Resistance: $R = V/I$

Capacitance: $C = Q/V$

Inductance: $L = \mathcal{N}\Phi/I$

Quadratic equations:

$$ax^2 + bx + c = 0 \quad \Rightarrow \quad x = \frac{1}{2a}\left(-b \pm \sqrt{b^2 - 4ac}\right)$$

Linear, first-order, homogeneous, differential equations:

$$\frac{dx}{dt} + \alpha x = 0 \quad \Rightarrow \quad x = x_0 e^{-\alpha t}$$

Linear, first-order, nonhomogeneous differential equations ($x_p = $ constant):

$$\frac{dx}{dt} + \alpha x = \alpha x_p \quad \Rightarrow \quad x = x_0 e^{-\alpha t} + x_p$$

l'Hôpital's rule:

$$\lim_{x \to a} f(x)/g(x) = \frac{df}{dx} \left/ \frac{dg}{dx}\right|_{x=a} \qquad (\text{if } f(a) = g(a) = 0)$$

Partial derivatives:

$$\delta f(x,y) = \frac{\partial f}{\partial x}\bigg|_y \delta x + \frac{\partial f}{\partial y}\bigg|_x \delta y$$

Average value:

$$\bar{x} = \frac{1}{T}\int_0^T x(t)\, dt$$

rms value:

$$x_{\text{rms}} = \sqrt{\frac{1}{T}\int_0^T x^2(t)\, dt}$$

Error analysis:

$$\delta f(x,y) = \sqrt{\left(\frac{\partial f}{\partial x}\delta x\right)^2 + \left(\frac{\partial f}{\partial y}\delta y\right)^2}$$

Taylor series:

$$f(x) = f(a) + \sum_{n=1}^{\infty} \frac{(x-a)^n}{n!} \frac{d^n f}{dx^n}(a), \qquad \text{where } n! = n(n-1)(n-2)\ldots 1$$

Fourier series:

$$f(t) = \sum_{n=-\infty}^{\infty} C_n e^{jn\omega_0 t} \qquad \text{where } C_n = \frac{1}{T} \int_{-T/2}^{T/2} f(t) e^{-jn\omega_0 t}\, dt$$

Fourier transform:

$$f(t) = \frac{1}{2\pi} \int_{-\infty}^{\infty} \bar{f}(\omega)\, e^{j\omega t}\, d\omega \qquad \text{where } \bar{f}(\omega) = \int_{-\infty}^{\infty} f(t) e^{-j\omega t}\, dt$$

F
Trigonometric Relations

$\sin \theta = -\sin (-\theta) = \cos (\pi/2 - \theta)$

$\cos \theta = \cos (-\theta) = \sin (\pi/2 - \theta)$

$\sin (\theta \pm \phi) = \sin \theta \cos \phi \pm \cos \theta \sin \phi$

$\cos (\theta \pm \phi) = \cos \theta \cos \phi \mp \sin \theta \sin \phi$

$\sin \theta \sin \phi = \frac{1}{2}[\cos (\theta - \phi) - \cos (\theta + \phi)]$

$\cos \theta \cos \phi = \frac{1}{2}[\cos (\theta + \phi) + \cos (\theta - \phi)]$

$\sin \theta \cos \phi = \frac{1}{2}[\sin (\theta + \phi) + \sin (\theta - \phi)]$

$\sin^2 \theta = \frac{1}{2}(1 - \cos 2\theta)$

$\cos^2 \theta = \frac{1}{2}(1 + \cos 2\theta)$

$\tan \theta = \sin \theta/\cos \theta$

$\cot \theta = 1/\tan \theta$

$\sec \theta = 1/\cos \theta$

$\operatorname{cosec} \theta = 1/\sin \theta$

$\sin^2 \theta + \cos^2 \theta = 1$

$A \cos \theta - B \sin \theta = \sqrt{A^2 + B^2} \cos (\theta + \phi)$ where $\phi = \tan^{-1} (B/A)$

G
Complex Numbers

$j = \sqrt{-1}$ $\qquad j^2 = -1$ $\qquad 1/j = -j$

$e^{\pi j/2} = j$ $\qquad e^{\pi j} = -1$ $\qquad e^{3\pi j/2} = -j$ $\qquad e^{2\pi j} = 1$

$e^{j\theta} = \cos\theta + j\sin\theta$

$\sin\theta = \dfrac{1}{2j}\left(e^{j\theta} - e^{-j\theta}\right)$

$\cos\theta = \tfrac{1}{2}\left(e^{j\theta} + e^{-j\theta}\right)$

$A + jB = \sqrt{A^2 + B^2}\, e^{j\phi}$ $\qquad$ where $\phi = \tan^{-1}(B/A)$

$\dfrac{1}{A + jB} = \dfrac{A - jB}{A^2 + B^2}$

$$d(fg) = f\,dg + g\,df$$

$$d(f/g) = \frac{g\,df - f\,dg}{g^2}$$

$$d(f^n) = nf^{n-1}\,df$$

$$d(e^{\alpha f}) = \alpha e^{\alpha f}\,df$$

$$d(\ln f) = df/f$$

$$d\sin\theta = \cos\theta\,d\theta$$

$$d\cos\theta = -\sin\theta\,d\theta$$

$$d\tan\theta = \frac{d\theta}{\cos^2\theta}$$

$$d\tan^{-1}f = \frac{df}{1+f^2}$$

Integrals

$$\int x^n dx = \frac{x^{n+1}}{n+1} + C$$

$$\int e^{\alpha x} dx = e^{\alpha x}/\alpha + C$$

$$\int \frac{1}{x} dx = \ln x + C$$

$$\int \ln x \, dx = x \ln x - x + C$$

$$\int \sin x \, dx = -\cos x + C$$

$$\int \cos x \, dx = \sin x + C$$

$$\int \tan x \, dx = -\ln \cos x + C$$

$$\int x \sin x \, dx = \sin x - x \cos x + C$$

$$\int x \cos x \, dx = \cos x + x \sin x + C$$

$$\int \sin x \cos x \, dx = \tfrac{1}{2} \sin^2 x + C$$

$$\int \sin^2 x \, dx = \tfrac{1}{2} x - \tfrac{1}{4} \sin 2x + C$$

$$\int \cos^2 x \, dx = \tfrac{1}{2} x + \tfrac{1}{4} \sin 2x + C$$

$$\int \sin mx \cos nx \, dx = -\frac{\cos (m-n)x}{2(m-n)} - \frac{\cos (m+n)x}{2(m+n)} + C \quad (m^2 \neq n^2)$$

$$\int_0^\pi \sin^2 x \, dx = \int_0^\pi \cos^2 x \, dx = \frac{\pi}{2}$$

$$\int_0^\pi \sin x \cos x \, dx = 0$$

$$\int_0^\infty e^{-ax^2} dx = \tfrac{1}{2} \sqrt{\frac{\pi}{a}}$$

$$\int_0^\pi e^{j(m-n)x} dx = \begin{cases} \pi & (m=n) \\ 0 & (m \neq n) \end{cases} \qquad m, n \text{ integers}$$

J
Approximations

The following formulas are valid for $|x| \ll 1$:

$(1+x)^n \simeq 1 + nx$

$\sin x \simeq x$

$\cos x \simeq 1 - x^2/2$

$\tan x \simeq x$

$\tan^{-1} x \simeq x$

$e^x \simeq 1 + x$

$a^x \simeq x \ln a$

$\ln(1+x) \simeq x$

K

Answers to Odd-Numbered Problems

Chapter 1

1.1 9.4×10^{-5} m/s

1.3   9.8 kW

1.5   800 Ω

1.7   0.833 Ω

1.9   1.618 R_0

1.11   5 V

1.13   0.5 A

1.15   1 V

1.17   12.5 V

1.19   (a) 1 V, (b) 1.625 V

1.21   40 A

1.23   Connect meters in series:
$V_1 = 880$ V, $V_2 = 220$ V

Chapter 2

2.1 $I_1 = I_2 + I_3$
$I_4 = I_3 + I_5$
$V_1 = I_1 R_1 + I_2 R_2$
$I_2 R_2 = I_3 R_3 + I_4 R_4$
$I_4 R_4 + V_2 + I_5 R_5 = 0$

2.3 $I_1 + I = I_2$
$V = I_1 R_1 + I_2 R_2$
$I_2 = \dfrac{V + I R_1}{R_1 + R_2}$

2.5   0.1 A

2.7 $V_T = 20$ V, $R_T = 150$ Ω,
$I_3 = 0.1$ A

2.9 $R = R_T$

2.11 $I_N = 1$ A, $R_N = 11$ Ω, $V_{oc} = 11$ V

2.13 $R_{11} = \dfrac{R_A R_B + R_B R_C + R_A R_C}{R_A + R_C}$
$R_{22} = \dfrac{R_A R_B + R_B R_C + R_A R_C}{R_A + R_B}$
$R_{12} = R_{21}$
$= \dfrac{R_A R_B + R_B R_C + R_A R_C}{R_A}$

2.15 $R_1 = 10$ Ω, $R_2 = 40$ Ω,
$R_3 = 10$ Ω

2.17 $\dfrac{R_1 R_2}{R_1 + R_2} + \dfrac{R_3 R_4}{R_3 + R_4}$

2.19 $V_T = 2.5$ V, $R_T = 13$ Ω

2.21 $I_{SC} = 23$ mA, $V_{OC} = 2.3$ V

Chapter 3

3.1   0.0177 μF

3.3   9.9 mH

3.5 $I(t) = \dfrac{V - V_0}{R} e^{-t/RC}$,
$V_C(t) = V + (V_0 - V) e^{-t/RC}$

3.7 $I(0) = \dfrac{V}{R_1 + R_2}$, $V_3(0) = 0$

$I(\infty) = \dfrac{V}{R_1 + R_3}$,

$V_3(\infty) = \dfrac{VR_3}{R_1 + R_3}$

3.9 $\dfrac{V}{R_1}(1 - e^{-R_1 R_2 t/(R_1 + R_2)L})$

3.11 $V_1(\infty) = V_2(\infty) = \dfrac{C_1 V_1(0)}{C_1 + C_2}$

3.13 $\dfrac{d^2 I}{dt^2} + \dfrac{1}{RC}\dfrac{dI}{dt} + \dfrac{1}{LC} I = \dfrac{V}{LRC}$

$I(0) = V/R, \dfrac{dI}{dt}(0) = -\dfrac{V}{R^2 C}$

3.15 $\dfrac{d^2 I_R}{dt^2} + \dfrac{1}{RC}\dfrac{dI_R}{dt} + \dfrac{1}{LC} I_R = \dfrac{V}{RLC}$

3.17 $V_C = 100\ t^2$, $V_L = 0.1$ V

Chapter 4

4.1 $I = \dfrac{\omega C V_0}{\omega^2 R^2 C^2 + 1}\left[\dfrac{1}{\omega RC} e^{-t/RC}\right.$

$\left. + \omega RC \cos \omega t - \sin \omega t\right]$

4.3 $I = \dfrac{\omega C V_0}{\sqrt{\omega^2 R^2 C^2 + 1}} \cos(\omega t + \phi)$,

where $\phi = \tan^{-1}(1/\omega RC)$

4.5 $Q = \omega_0 L/R$

4.7 $\phi = 154.3^0$

4.9 $V_T = 3.54\ e^{100 jt + j\pi/4}$

$Z_T = 50 + 50 j$, $R_T = 50\ \Omega$,

$L_T = 0.5$ H

4.11 $|V_{out}/V_{in}|$

$= \dfrac{1}{\sqrt{1 + \left[\dfrac{\omega L}{R(\omega^2 LC - 1)}\right]^2}}$

$\phi = \tan^{-1} \dfrac{\omega L}{R(\omega^2 LC - 1)}$

4.13 $\omega_C = \dfrac{R + R_L}{RR_L C}$

4.15 16 Hz to 8 MHz

4.17 $\omega^2 = \dfrac{1}{R_3 C_3 R_4 C_4}$,

$\dfrac{R_2 R_3}{C_3} = \dfrac{R_1 R_4}{C_3} + \dfrac{R_1 R_3}{C_4}$

4.19 $|V_{out}/V_{in}| = \dfrac{R_2}{R_1 + R_2}$

(independent of ω)

4.21 $L = 40$ mH, $I_{rms} = 7.07$ A

4.23 $L_P = 4$ H, $k = 0.998$

Chapter 5

5.1 $V(t) = \dfrac{2V_0}{\pi}\displaystyle\sum_{n=1}^{\infty}\dfrac{1}{n}\left[1 - \cos\dfrac{n\pi}{2}\right]$

$\sin n\omega_0 t$

5.3 $V(t) = \dfrac{2V_0}{\pi} - \dfrac{4V_0}{\pi}\displaystyle\sum_{\substack{n=2 \\ n\ even}}^{\infty}$

$\dfrac{\cos n\omega_0 t}{n^2 - 1}$

5.5 $C_n = \tfrac{1}{2}(a_b - jb_n)$

5.7 $V_{out} \simeq -8.1 \times 10^{-4}\ V_0 \displaystyle\sum_{\substack{n=1 \\ n\ odd}}^{\infty}$

$\dfrac{\sin n\omega_0 t}{n^3}$

5.9 $\quad |\bar{V}(\omega)| = \dfrac{2V_0}{\omega}(1 - \cos \omega\tau)$

5.11 $\quad I(t) = \dfrac{V_0 C}{\pi} \displaystyle\int_{-\infty}^{\infty}$

$\dfrac{\omega RC + j(1 - \omega^2 LC)}{\omega^2 R^2 C^2 + (1 - \omega^2 LC)^2}$

$\sin(\omega\tau/2) e^{j\omega t}\, d\omega$

5.13 $\quad \phi = -51°$

5.15 $\quad R = \mathcal{Z}_0^2/R_L,\ C = L_L/\mathcal{Z}_0^2$

5.17 $\quad C' = \varepsilon w/d,\ L' = \mu d/w,$

$\mathcal{Z}_0 = \sqrt{\mu/\varepsilon}\, d/w$

5.19 $\quad v_g = \dfrac{v_p}{1 + (\lambda/2\omega)^2} < v_p < c$

Chapter 6

6.1 $\quad I = 10.2$ mA, $V = 54$ V

6.3 $\quad$ 0.48 to 0.66 V

6.5 $\quad$ 0.25 W (half-wave),
0.5 W (bridge)

6.7 $\quad$ Open: looks like a half-wave
rectifier.
Short: draws large current from
source.

6.9 $\quad I_{surge}/I_{av} = 2\sqrt{\pi\omega RC}$

6.11 $\quad$ 20%

6.13 $\quad V_R(dc) = 6.37$ V,
$L = 17.7$ H

6.17 $\quad T = 15.7$ ms, $I_{max} = 1$ A

6.19 $\quad P_S = 4$ W, $P_R = 2$ W, $P_L = 1$ W,
$P_Z = 1$ W

6.21 $\quad$ 4%

Chapter 7

7.1 $\quad V_{PC} = 82$ V, $I_P = 27$ mA

7.3 $\quad$ (a) increases, (b) decreases,
(c) decreases

7.5 $\quad R_P \simeq 10$ kΩ, $R_C = 50$ Ω,
$R_G = 100$ kΩ,
$A \simeq 50,\ C_G = 0.016\ \mu$F

7.7 $\quad A = -\dfrac{\mu R_P}{R_P + r_p + (1 + \mu)R_C}$

$\simeq -\dfrac{R_P}{R_C}$ (for μ large)

7.9 $\quad C_{in} = (1 + |A|)C$

7.11 $\quad A = 0.46$

7.13 $\quad$ Common cathode: $A \simeq -100$,
$R_{in} = 1$ MΩ, $R_{out} \simeq 10\ k\Omega$
Cathode follower: $A \simeq 0.5$,
$R_{in} = 1$ MΩ, $R_{out} \simeq 50$ Ω
Grounded grid: $A \simeq 100$,
$R_{in} \simeq 50$ Ω, $R_{out} \simeq 10$ kΩ

7.15 $\quad$ Peaks of wave are flattened

7.17 $\quad R_G = 1$ MΩ, $R_D = 1$ kΩ,
$R_S = 100$ Ω

7.19 $\quad A_1 = 0.995,\ A_2 = -0.995$

7.21 $\quad$ Common source: $A = -g_{fs}R_D$,
$R_{in} = R_G,\ R_{out} = R_D$
Source follower: $A = 1$,
$R_{in} = R_G$

$R_{out} = \dfrac{R_S}{g_{fs}R_S + 1}$

Grounded gate: $A = g_{fs}R_D$,

$R_{in} = \dfrac{R_S}{g_{fs}R_S + 1},\ R_{out} = R_D$

Chapter 8

8.1 $V_{CE} = 11$ V, $I_C = 4.5$ mA

8.3 $V_{CE} = 0.5$ V, $I_C = 9.5$ mA

8.5 $r_{tr} = \dfrac{r_B}{1 + \beta} + r_E$

8.7 $V_B = 2$ V, $V_E = 1.4$ V, $V_C = 3$ V,
      $I_B = 14$ μA, $I_E = I_C = 1.4$ mA

8.9 $A = -3$

8.11 $V_{out} = 4 - 0.99 \sin \omega t$

8.13 $C_1 = 6$ μF, $C_2 = 123$ μF

8.15 $R_{in} = 75$ Ω, $A = 49.5$

8.17 $I_L = 1$ A, $P_L = 10$ W, $P_T = 10$ W,
      $P_R = 0.25$ W, $P_z = 0$ W

8.19 $CMRR = 100$

Chapter 9

9.1 $A = -(R_f + R_2 + R_2 R_f / R_1)/R_i$

9.3 $R_{in} = 5 \times 10^{10}$ Ω

9.5 $R_1 = R_2 = 1000$ Ω, $R_3 = 2000$ Ω

9.7 $A = \dfrac{R_1}{R_1 + R_2}$

9.9 $A = 1$

9.11 $R_{in} = -R_1 R_3 / R_2$

9.13 $R_1 C = 1/7$, $R_2 C = 4$

9.17 $V_{out} = -\dfrac{kT}{e} \ln (V_{in}/\beta I_0 R)$

9.19 $V_{rms} = 0.513$ μV

Chapter 10

10.1 $R = 10$ MΩ

10.3 $A_0 = 29$

10.5 $R_f / R_i = 1$

10.7 $V_{in} = 6$ V

10.11 $T = 4$ ms

10.13 $V = V_0 e^{-t/2RC} \sin \omega t$

10.15 $\beta_1 \beta_2 > 1$

10.17 $V(t) = \dfrac{V_0}{2\pi} + V_0 \cos \omega_0 t$

         $+ \dfrac{V_0}{2\pi} \sin \omega_0 t$

10.19 $V_{out} \simeq 100 (1 - e^{-1000/f})$

Chapter 11

11.1   107, 2018, 2671

11.3 10010101_2, 95_{16},
      11000110000_2, 630_{16}

11.5 -354

11.7 $V_0 = (A + \bar{B}) \cdot \bar{C}$

11.19   1011, 0101, 0010, 1001, 0100

11.21   6.25%, 0.0015%

Chapter 12

12.1   280 m, 3.38 m, 43 cm, 11 m,
      12 cm

12.3   164 MHz

12.5   62.3 km, 1000 ft

12.7 V_0 at ω_0, $V_0/2$ at $\omega_0 \pm \Delta\omega$

12.9 $V = 10 \sin 2\pi (10^8 + 10^4$
      $\sin 6283 t) t$

12.11   80–1250 kHz; 910, 682.5,
       606.7, 568.8, 546 kHz;
       910 kHz

12.13   1.19 km

12.15 10^4, 300 m, 300 km, 1 MHz

12.17   4.9 kHz

L
Laboratory Experiments

Included here are fourteen suggested laboratory experiments to accompany the text. The experiments are each designed to be completed in a single three-hour laboratory period.

It is suggested that students work in pairs at benches provided with the following basic equipment:

Dual power supply, current-limited, regulated (0–20 V)

Volt-ohm-milliammeter (20 kΩ/V)

High-impedance voltmeter ($\geq$10 MΩ)

Function generator, sine, square, triangular waves (10 Hz–500 kHz)

Oscilloscope (dc–1 MHz)

In addition, an assortment of resistor, capacitor, and inductor substitution boxes or individual components are required. Several of the experiments are facilitated by having preassembled circuit boards containing most of the required components.

Students are encouraged to write up the results of the experiments with some care. The wise student will do enough for the analysis during the laboratory period to ensure that the experiment was done properly. Wherever possible, the measured values of quantities should be compared with theoretical calculations and the percentage error given. When plotting graphs, make sure the axes are adequately labeled with the quantity and its units, and put the measurements on the graph as discrete data points and the theory, where appropriate, as a solid line. It is a good idea to write a short, one-paragraph summary at the end of the report, stating what was learned from the experiment.

D.C. Measurements

The purpose of this experiment is to acquaint the student with the various instruments available in the laboratory for measuring dc currents and voltages and to acquaint the student with the limitations of these devices.

Apparatus required: 0–20 V power supply, two resistor substitution boxes, VOM, VTVM (or equivalent), 24-V incandescent lamp.

Procedure:

1. Set up the voltage divider shown in figure 1.7 using R_1 and R_2 of approximately 100 Ω. Test the voltage divider relation using the VOM and again using the VTVM.

2. Repeat the above measurements using 1 MΩ for R_1 and R_2. When does a voltmeter give a useful measurement in this circuit?

3. Measure at least two resistances, one about 100 Ω and one about 50 kΩ, using each of the circuits below:

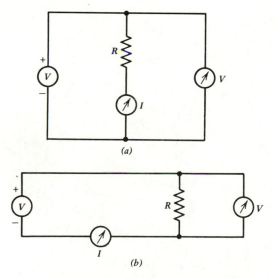

(a)

(b)

For each case, correct the measured resistance for the finite internal resistance of the meters. Explain under what circumstances each circuit is best for such resistance measurements.

4. By varying the power supply voltage, measure and plot the voltage across the resistance as a function of current through the resistance in one of the best circuits. Determine R from the slope of the line. Estimate the percentage accuracy of the measurement.

5. Measure and plot the voltage across an incandescent lamp as a function of current through the lamp. Is the lamp a linear component?

Circuit Theorems

This experiment illustrates the superposition theorem, Thevenin's theorem, and the reciprocity theorem for linear circuits; and the failure of these theorems when the circuit contains nonlinear components.

Apparatus required: Dual 0–20 V power supply, VOM, VTVM, 3-port resistor network with switch to add an incandescent lamp.

Procedure:

1. With the switch in the R position, apply various voltages V_2 at port 2 and simultaneously apply V_3 at port 3 and measure the short circuit current at port 1. Does the superposition theorem hold?

2. With the switch in the R position, apply a voltage V_1 of 10 V at port 1 and determine the open circuit voltage V_2 and the short-circuit current I_2 with terminals 3 short circuited. Find the Thevenin and Norton parameters for port 2.

3. With the switch in the R position, place a voltage source V_1 at port 1 and an ammeter at port 2 with terminals 3 short circuited. Now interchange the voltage source and the ammeter and repeat the measurements. Is the reciprocity theorem obeyed?

4. With the switch in the R position, measure all 9 R parameters for the circuit ($R_{ij} = V_j/I_i$), remembering to short-circuit the unused terminal pair(s).

5. With the switch in the L position, repeat as many of the above measurements as necessary to convince yourself that the circuit theorems are *not* obeyed.

6. Make a diagram of the circuit. Analyze and display the results of the data and a comparison with the values calculated from the diagram for V_T, R_T, I_N, R_{11}, R_{12}, and so on.

Wheatstone Bridge

This experiment is to familiarize the student with the Wheatstone bridge and its ability to make very precise measurements using the null method.

Apparatus required: 0–20 volt power supply, sensitive galvanometer, 0.5% precision 1000-Ω resistor, 0.05% precision resistor decade box.

Procedure:

1. Set up a Wheatsone bridge circuit as shown in figure 2.13. Use a precision (0.05%) resistor decade box set to 1000 Ω for R_3. Use a precision (0.05%), 1000-Ω resistor for R_4. Use low-precision, 1000-Ω resistors for R_1 and R_2. Since R_1 and R_2 will not be precisely equal, it will be necessary to place high-value resistors in parallel with either R_1 or R_2 in order to achieve a balance. Once this is done, verify that a 1-Ω change in R_3 either above or below 1000 Ω will produce a noticeable reading on the galvanometer.

2. Use the already-balanced bridge with R_4 changed to an unknown resistor to measure by direct reading of R_3 the correct resistance of resistors of about 500 Ω, 1000 Ω, and 2000 Ω, with an accuracy of approximately 0.1%. Repeat until the results agree with those obtained by your instructor.

Oscilloscope

The purpose of this experiment is to permit the student to become familiar with the operation of the cathode ray oscilloscope.

Apparatus required: Oscilloscope, function generator, 0–20 V power supply, VOM, resistor substitution box, 6.3-V transformer.

Procedure:

1. With the function generator connected to the vertical input of the oscilloscope, observe sine, square, and triangular waves of several different frequencies.

2. Check the sweep-speed calibration of the oscilloscope by applying a 60-Hz sinusoidal voltage from the 6.3-V transformer to the vertical input.

3. Using a dc voltage source in conjunction with a VOM, check the voltage calibration of the oscilloscope.

4. Apply a 60-Hz voltage from the 6.3-V transformer to the horizontal input. Use the function generator to apply a sinusoidal ac voltage to the vertical input. Observe the Lissajous figure that results when the function generator is carefully adjusted to a frequency that is in a rational ratio to 60 Hz (i.e., 20 Hz, 30 Hz, 40 Hz, 60 Hz, 120 Hz, etc.). How accurately calibrated is the function generator?

5. Devise and execute an experiment to determine the input resistance of the oscilloscope.

Transient Series RLC Circuit

The purpose of this experiment is to permit the student to study the transient response of a series RLC circuit in the overdamped, critically damped, and underdamped cases.

Apparatus required: 0–20 V power supply, mercury relay, oscilloscope, resistor , inductor, and capacitor substitution boxes.

Procedure:

1. Construct the following circuit:

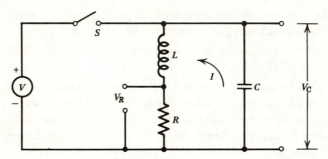

The switch S is a mercury relay that is driven by the 60-Hz voltage from the power line. Verify that it opens and closes every 1/120 s. Adjust the oscilloscope so that it triggers on the opening of the switch. The behavior of the circuit will be studied just after the switch opens, at which time the source is disconnected and the circuit consists only of a series RLC.

2. What are the initial conditions for I, V_c, and dI/dt?

3. Choose R, L, and C such that the circuit is underdamped with a Q of about 20. Measure the frequency of oscillation and compare with equation 3.23.

4. Measure the damping rate of the oscillation and compare with equation 3.25.

5. Increase R until the circuit is critically damped, and compare with the value predicted for R.

6. Increase R further to produce strongly overdamped behavior and compare the rise and fall times of the current with the expected values.

Filter Circuits

The purpose of this experiment is to acquaint the student with the measurement of phase and amplitude for various types of filter circuits with sinusoidal sources.

Apparatus required: sine wave generator, oscilloscope, resistor, inductor, and capacitor substitution boxes.

Procedure:

1. Set up and measure the attenuation and phase of a low-pass RL filter as shown in figure 4.8(a) as a function of frequency.

2. Set up and measure the attenuation and phase of a high-pass RC filter as shown in figure 4.9(a) as a function of frequency.

3. Set up and measure the attenuation and phase of a resonant filter as shown in problem 4.10 as a function of frequency. Choose R, L, and C to give resonance at 5000 Hz with a Q of 10.

For each of the above cases, compare the results with the theoretical predictions. A programmable calculator or computer should prove very helpful here.

Fourier Series

The purpose of this experiment is to demonstrate that a periodic waveform can be decomposed into an infinite sum of sine waves with harmonically related frequencies.

Apparatus required: function generator, oscilloscope, 5-Ω resistor, 100-Ω resistor, 100-mH inductor, and capacitor decade box.

Procedure:

1. Set up the following spectrum analyzer circuit:

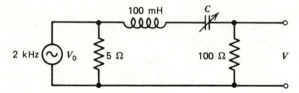

The 5-Ω resistor lowers the Thevenin equivalent resistance of the 2-kHz square wave source and raises the Q of the RLC circuit.

2. By adjusting C with the oscillator fixed at 2 kHz, measure the magnitude of each Fourier component up to at least $n = 9$. Record the capacitance required for resonance at each harmonic. It may be helpful to vary the oscillator frequency slightly (a few percent) to maximize V if the resonance falls between switch positions of the capacitor decade box.

3. Plot the ratio $|V_n/V_0|$ versus n on log-log graph paper and compare the results with the theoretical values predicted for the Fourier series representation of figure 5.4(a).

4. Repeat steps 1–3 above using triangular rather than square waves.

Characteristics of Ge, Si, and Zener Diodes

The purpose of this experiment is to enable the student to examine and understand the properties of germanium and silicon solid state diodes and Zener diodes.

Apparatus required: 0–20 V power supply, resistor substitution box, VOM, VTVM, diodes (Ge, Si, Zener), 6.3-V transformer, and oscilloscope.

Procedure:

1. Set up the following circuit.

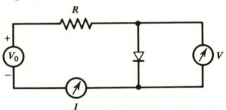

By varying V_0 and R, measure the V-I characteristic of a germanium diode from $V_0 = -10$ V up to whatever positive voltage is necessary to cause 10 mA to flow through the diode. Note that either the voltmeter must have a very high internal resistance or a correction to the measured current must be made.

2. Compare the measured results with equation 6.4 and estimate the values of I_o, kT/e, and r_{ohmic}.

3. Repeat steps 1 and 2 for a silicon diode, using forward currents up to 100 mA.

4. Set up the following circuit to display the I-V characteristic of the diodes on an oscilloscope:

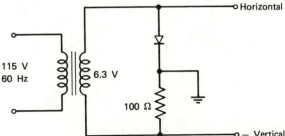

What are the approximate forward voltage drops for the two diodes?

5. Display using the circuit above the *I-V* characteristic of a Zener diode. Determine the breakdown voltage, and decide whether the diode is germanium or silicon.

Rectifier Circuits

The purpose of this experiment is to permit the student to examine the properties of various rectifier circuits and to examine the influence of a filter on the output of the rectifier.

Apparatus required: 6.3 VCT transformer, four silicon diodes, 10-μF capacitor, resistor substitution box, and oscilloscope.

Procedure:

1.   Set up the half-wave rectifier shown in figure 6.6, and compare the output waveform with the predicted value. Be sure to measure the output voltage of the transformer, since it is likely to be greater than 6.3 V rms when the output current is small. Also consider the forward voltage drop of the diode.

2.   Set up the full-wave rectifier shown in figure 6.7, and compare the output waveform with the predicted value.

3.   Set up the bridge rectifier shown in figure 6.8, and compare the output waveform with the predicted value.

4.   Using a 10-μF capacitor, connect a capacitive filter as shown in figure 6.9 to each of the rectifier circuits and measure the dc output voltage and ripple for several values of load resistance. Compare the observed ripple with that predicted by equation 6.7.

FET and Bipolar Transistors

The purpose of this experiment is to acquaint the student with the properties of field effect and bipolar transistors.

Apparatus required: Dual 0–20 V power supply, VOM, VTVM, resistance substitution box, p-channel FET, and npn bipolar transistor.

Procedure:

1. Set up the following circuit:

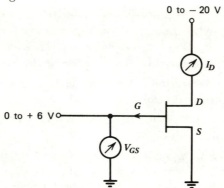

Measure and plot the drain current I_D versus drain-to-source voltage V_{DS} for several gate-to-source voltages V_{GS} in the range 0–6 V.

2. Identify on your graph the ohmic region, the pinch-off region, and the break-down region, and estimate the value of the output resistance r_{os} in the pinch-off region.

3. Set up the following circuit:

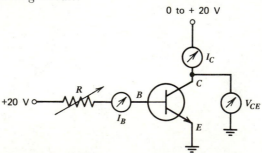

Measure and plot the collector current I_C versus collector-to-emitter voltage V_{CE} for several values of base current I_B in the range 0–100 μA.

4. From the graph estimate the value of beta for the transistor.

5. Heat or cool the transistor, and note the effect of temperature on beta.

6. Measure the base-to-emitter voltage as a function of base current, and verify that the base-to-emitter junction behaves like a forward-biased diode. Is the transistor germanium or silicon?

Common Emitter Amplifier

The purpose of this experiment is to teach the student how to design a simple transistor amplifier.

Apparatus required: 0–20 V power supply, silicon *npn* transistor, assorted resistors and capacitors, sine wave generator, and oscilloscope.

Procedure:

1. Design a common emitter amplifier as shown in figure 8.6 with the following characteristics: $V_{cc} = 15$ V, $A = -25$, $R_{out} = 50$ kΩ, lower cutoff frequency $f = 100$ Hz.

2. Construct the circuit, and determine that the operating point is correct. It may be necessary to vary R_1 somewhat from your design value to achieve proper operation.

3. Apply a 1-kHz sine wave at v_{in}, and observe the saturation and cutoff that occurs at the output when the input amplitude is too large.

4. Using a suitably small value of v_{in}, measure the amplification, input resistance, and output resistance, and compare with the predicted values.

5. Measure the amplification as a function of frequency from 20 Hz to at least 500 kHz, and plot the results on log-log paper.

6. With a suitable capacitor in parallel with R_E, repeat step 5, and plot the results on the same graph. Does the measured amplification agree with the prediction based on a reasonable estimate of the transresistance?

Linear Op Amp Circuits

The purpose of this experiment is to permit the student to set up and examine the use of op amps to perform the four basic linear mathematical operations of addition, subtraction, integration, and differentiation.

Apparatus required: Dual 0–20 V power supply, operational amplifier, assorted resistors and capacitors, function generator, 1.5-V battery, and oscilloscope.

Procedure:

1. Set up the circuit in figure 9.9(a) for adding two voltages. Use resistors in the range 10–100 kΩ. With $V_2 = 0$, select resistors so that $V_{out} = -3V_1$. Using a sinusoidal input for V_1, measure the amplification of the system at low frequency and small amplitude.

2. Determine the value of V_{out} at which saturation occurs.

3. Measure and graph the amplification as a function of frequency.

4. Using a sinusoidal voltage for V_1 with an amplitude such that V_{out} is close to saturation, increase the frequency until the slew rate is observed. How does the observed slew rate compare with the published specifications for the op amp?

5. Using a sinusoidal voltage for V_1 and a 1.5-V battery for V_2, verify that the output is given by equation 9.14.

6. Set up the circuit in figure 9.9(b) for subtracting two voltages. Using a sinusoidal voltage for V_1 and a 1.5-V battery for V_2, verify that the output is given by equation 9.16.

7. Set up the circuit in figure 9.9(c) for integrating a voltage. It may be necessary to place a high-value resistor (1–10 MΩ) in parallel with the capacitor to prevent the input offset voltage from saturating the op amp. Using a square wave for V_{in}, verify that the output is given by equation 9.18.

8. Set up the circuit in figure 9.9(d) for differentiating a voltage. Using a triangular wave for V_{in}, verify that the output is given by equation 9.19.

Nonlinear Op Amp Circuits

The purpose of this experiment is to acquaint the student with some of the nonlinear applications of op amps.

Apparatus required: Dual 0–20 V power supply, operational amplifier, assorted resistors, capacitors, and inductors, silicon diode, function generator, 1.5-V battery, and oscilloscope.

Procedure:

1. Set up the comparator circuit shown in figure 9.13(a). Set V_1 at 1.5 V. Use a low-frequency sine wave with variable amplitude for V_2. Verify that the circuit works as a comparator. How would one use an op amp to generate a symmetric square wave, starting with a sine wave at its input?

2. Set up the latch circuit shown in figure 9.13(b). Show that the circuit remembers the last polarity of V_{in}.

3. Set up the logarithmic amplifier shown in figure 9.11(a). Sketch the shape of V_{out} when V_{in} is a triangular wave.

4. Set up an exponential amplifier as shown in figure 9.11(b). Sketch the shape of V_{out} when V_{in} is a triangular wave.

5. Construct at least one oscillator circuit using positive feedback. Measure its frequency of oscillation and compare the result with the expected value.

Digital Circuits

The purpose of this experiment is to introduce the student to the fundamentals of digital circuits.

Apparatus required: 5-V power supply, circuit board containing TTL: 8 two-input NAND gates, 8 JK flip-flops, 4 toggle switches, and 16 indicator lamps.

Procedure:

1. Verify the truth table for the NAND gate in figure 11.4(*b*).

2. Use NAND gates to construct the following circuits and verify their operation: 2-input AND gate, 2-input OR gate, and 2-input NOR gate.

3. Cross couple two NAND gates to make an RS flip-flop similar to that in figure 11.10.

4. Make a clocked RS flip-flop and a D-flip-flop as shown in figure 11.11.

5. Verify the truth table for the JK flip-flop in figure 11.12(*b*).

6. Construct an 8-bit counter as shown in figure 11.13(*a*). Show that it acts as a frequency divider.

7. Construct an 8-bit shift-register as shown in figure 11.13(*b*). Connect the output to the input and make a circular shift-register.

Index